Molecular Toxicology

Molecular Toxicology

Nick Plant
School of Biomedical and Life Sciences, University of Surrey, Guildford, UK

Taylor & Francis
Taylor & Francis Group

LONDON AND NEW YORK

© Taylor & Francis Publishers, 2003

First published 2003. Transfered to digital print 2009.

A CIP catalogue record for this book is available from the British Library.

ISBN 1 85996 345 5

Taylor & Francis Publishers
2 Park Square, Milton Park, Abingdon, OX14 4RN, UK
270 Madison Avenue, New York, NY 10016, USA

Taylor & Francis Publishers is a member of the Taylor & Francis Group.

Production Editor: Andrea Bosher
Typeset by Charon Tec Pvt. Ltd, Chennai, India

Cover photograph of deadly nightshade kindly supplied by Professor Alastair Fitter

Contents

Abbreviations

ABP	4-amino biphenyl
AFB	aflatoxin B1
AhR	aryl hydrocarbon receptor
ARE	antioxidant response element
ARNT	AhR nuclear translocator
AUC	area under the curve
BER	base excision repair
BLAST	basic local algorithm search tool
CNS	central nervous system
CS-lyase	cysteine conjugate β-lyase
CYP	cytochrome P450
DCM	dichloromethane
dsDNA	double strand DNA
EpRE	electrophile responsive element
ER	endoplasmic reticulum
ES cells	embryonic stem cells
EST	expressed sequence tag
FMO	flavin mono-oxygenase
GRα	glucocorticoid receptor
GRAS	generally recognized as safe
GST	glutathione-S-transferase
HAT	histone acetyl transferase
HCA	heterocyclic aromatic amine
HI	Hazard index
HSP	heat shock protein
IAP	inhibitor of apoptosis
LCR	locus control region
MALDI TOF	matrix-assisted laser desorption ionization time-of-flight MS
MAPK	mitogen activated protein kinase
MAPKK	MAPK kinase
MAPKKK	MAPKK kinase
MDR	multidrug resistance protein
MPP^+	1-methyl-4-phenylpyridium
MPPP	1-methyl-4-phenyl-4-propionoxypiperidine
MPTP	1-methyl-4-phenyl-1,2,3,6-tetrahydropyridine
mRNA	messenger RNA
MRP	multidrug resistance related protein
MS	mass spectroscopy
MT	metallothionein
NAPQI	N-acetyl-ρ-benzoquinoneimine
NCBI	National Centre for Biotechnology Information

NER	nucleotide excision repair
NMDA	*N*-methyl-D-aspartate receptor
NNTI	non-nucleoside reverse transcriptase inhibitor
NOS	nitric oxide synthase
NSAID	non-steroidal anti-inflammatory drugs
NTE	neuropathy targeted esterase
OAT	organic anion transporter
OATP	organic anion transport protein
OCT	organic cation transporter
PAGE	polyacrylamide gel electrophoresis
PAH	polycyclic aromatic hydrocarbon
PAPS	3'-phosphadenosine-5'-phosphosulphate
PCB	polychlorinated biphenyl
PCR	polymerase chain reaction
P-gp	P-glycoprotein
pI	isoelectric point
PNS	peripheral nervous system
PP	peroxisome proliferator
PPAR	peroxisome proliferator activated receptor
PVC	poly vinyl chloride
PXR	pregnane X receptor
Redox	reduction:oxidation
ROS	reactive oxygen species
RT-PCR	reverse transcriptase PCR
SELDI	surface-enhanced laser desorption/ionization MS
SNP	single nucleotide polymorphism
SSCP	single strand conformational polymorphism
ssDNA	single strand DNA
SULT	sulphotransferase
TBP	TATAA binding protein
TEF	toxic equivalency factor
TMA	trimethylamine
UGT	UDP-glucoronosyltransferase
UTR	untranslated region
WoE	weight of evidence
WWW	World Wide Web (The Internet)
XRE	xenobiotic response element

Preface

Ever since Neanderthal man ate some shiny red berries and then promptly keeled over the science of toxicology has been inextricably linked with the development of mankind. Whether we are aware of it or not we are exposed to thousands of chemicals every day, and many of these are potentially harmful to us. It is a testament to the versatility of the body that we do not notice such exposures; the body instead has many systems to remove chemicals before they can exert any toxic effects on us. The role of the toxicologist is to understand these reactions, allowing a better prediction of the effects of exposure to toxins, and the production of new chemicals/drugs that do not possess potentially harmful properties.

Each generation of toxicologist has used new technologies to delve deeper into the mechanisms of toxicity. This text will focus on the latest wave of 'new' technology, the use of molecular measurements to further understand the response of the body toxic insult – this is the science of Molecular Toxicology.

This book would not have been possible without the input of many people. Ideas, comments and enthusiasm from colleagues have helped guide my thoughts, while my PhD students have put up stoically with a distracted supervisor during the writing process. However, the deepest thanks must go to my wife Kate, for her comments, proofreading and most importantly patience and enthusiasm in equal measures. Without whom… .

General concepts in toxicology

1.1 What is toxicology?

Before we can begin to examine how the body responds to a toxic insult we must first ask 'What is a toxic insult?', and indeed 'What is toxicology?'. The Dictionary of Toxicology defines toxicology as 'The science that deals with poisons (toxicants) and their effects' (Hodgson *et al.*, 1998). It further defines a poison as 'any substance that causes a harmful effect, either by accident or design, when administered to a living organism'. From this we can define toxicology as the study of chemicals that harm living organisms. Such a definition would suggest that toxicology is an applied science, i.e. its study has a direct, measurable impact upon society. What impact does it therefore have? *Table 1.1* has a list of potential roles for toxicologists in everyday life. The wide range of job roles shown again underlines the importance of the toxicologist to mankind in general; ranging from protecting our environment to the production of safer drugs with fewer side effects.

To begin to understand how a chemical may damage the body (i.e. be toxic to it) we must first understand how the chemical will interact with the body.

1.2 Passage of a chemical through the body

At the simplest level, interaction of a chemical with the body can occur at the point of contact; most commonly through the skin. For example, if you spill acid on your hand then the corrosive damage is caused *at that site* by the interaction of the acid with the skin. This is a very simple case, affecting only a tiny portion of the body. For the majority of exposures however, the chemical enters into the body, and then exerts its toxic effects at several sites

Table 1.1 Roles of the toxicologist

Role	Possible duties
Industrial toxicologist	Development of 'safe' drugs, agrichemicals etc.
University toxicologist	Education and research of toxicity
Clinical toxicologist	Specialist on the toxic effects of chemicals on man
Forensic toxicologist	Investigation of the role of toxic chemicals in legal cases
Eco toxicologist	Effects of toxic chemicals on the eco system
Regulatory toxicologist	Advice and regulation of chemicals that are toxic
Occupational toxicologist	Potential effects of toxic chemicals in 'everyday' use

within the body. This, as we will see, also allows the body to mount a bio-chemical defence against the toxic insult, trying to stop it being toxic to the body through both chemical modification (detoxification) and rapid removal from the body. Therefore, to understand the body's defence mech-anisms against a toxic insult we should first consider the passage of any chemical through the body; this occurs in a number of semi-independent phases. Firstly, the chemical must get into the bloodstream of the body (absorption), and from there it can move around the body to the various organs (distribution). It can then be chemically altered (metabolism) to increase its rate of removal from the body (excretion). This process is often referred to by the initial letters of each of these phases (i.e. ADME).

Detailed discussion of all of the phases of ADME, and their roles in drug metabolism, are given by texts such as Gibson and Skett (2001), and the interested reader is directed towards these. While a full description of ADME is outside the scope of this text, a brief introduction is given below.

1.2.1 Absorption

The most obvious routes of exposure for the body to a chemical are via eat-ing (ingestion), breathing (inhalation) or skin (dermal absorption). To enter the body via any of these routes the chemical must first cross a cell mem-brane to enter the plasma; this can be achieved either via passive diffusion, facilitated diffusion or active transport (*Figure 1.1*).

Of these mechanisms the simplest is passive diffusion, as all that is required is a non-polar, lipophilic compound (increasing its ability to cross a lipid membrane) and a concentration gradient. Hence, the majority of chemicals

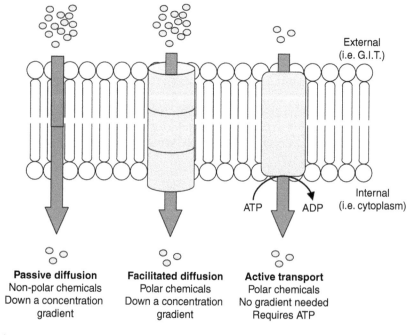

Passive diffusion	Facilitated diffusion	Active transport
Non-polar chemicals	Polar chemicals	Polar chemicals
Down a concentration gradient	Down a concentration gradient	No gradient needed
		Requires ATP

Figure 1.1

Mechanisms of absorption.

undergo absorption via this method. Facilitated diffusion and active transport allow the transport of chemicals across membranes that are not good candidates for absorption by passive diffusion. This may be because the chemicals are polar (and hence will not easily enter a non-polar, lipid membrane) or are not present at a sufficiently high concentration to form a good concentration gradient. In facilitated diffusion, 'protein pores' allow the transfer of polar molecules across a membrane, although a concentration gradient is still required. In active transport, however, movement of polar chemicals can occur against a concentration gradient, but at the expense of energy (in the form of ATP $\rightarrow$ ADP). As both facilitated diffusion and active transport utilize membrane proteins the abundance of these proteins can be a limiting factor in the rate of absorption. If a chemical is at a very high concentration then its rate of transfer across the membrane may be determined by the number of proteins that can carry it, as opposed to the chemico-physical characteristics of the chemical.

1.2.2 Distribution

Once a chemical is in the plasma it can be distributed around the body, thus explaining the body-wide effects of some chemicals. However, blood flow is not equal to all organs and therefore neither is the distribution of a chemical in the bloodstream. In general the major body organs (i.e. liver, heart, lungs, etc.) receive the majority of cardiac output, followed by tissues such as the muscles. Finally tendons, teeth and ligaments receive very little blood flow and therefore only a small percentage of any chemical entering the body will end up in these tissues. *Table 1.2* shows the relative blood flows to different body compartments.

Of the well-perfused body organs described above, the liver deserves special mention. This organ is the site for the majority of metabolism within the body, and has the highest concentration of drug-metabolizing enzymes present in the body. The high blood flow to the liver ensures that most chemicals are taken up very efficiently into the liver, thus ensuring their rapid metabolism. While the liver is thought of as the 'seat of metabolism' it should be remembered, however, that all tissues in the body have some drug-metabolizing enzymes and are thus capable of some metabolism; the liver merely has the highest concentration and widest variety of these enzymes. As we will see in Chapters 2 and 3, metabolism plays a vital role in both detoxification and toxication pathways for many chemicals, and hence the liver is a key site for studying chemical toxicity.

1.2.3 Metabolism

Metabolism is defined as 'the total of all chemical transformations of normal body constituents taking place in a living organism, whether they are

Table 1.2   Blood flow to body compartments

Degree of perfusion	Examples	Blood flow
Good	Lungs, liver, heart	20–400 ml/min/100 g
Medium	Muscle, skin	1–20 ml/min/100 g
Poor	Teeth, ligaments, tendons	<1 ml/min/100 g

synthetic (anabolic) or degradative (catabolic) reactions' (Hodgson *et al.*, 1998). Thus, the body uses chemicals within it to provide important base units for the formation of other chemicals, including energy in the form of ATP. In addition, this process helps to regulate levels of chemicals so that their temporal effects on the body can be regulated. Finally, the prevention of a build-up of chemicals prevents the formation of levels that might lead to adverse (i.e. toxic) effects of these chemicals.

This process, however, does not only apply to endogenous compounds; any compound which enters the body from outside (i.e. exogenous) is also subject to metabolism. Indeed, in relation to toxicity it is probably more apposite to discuss such xenobiotic metabolism, which can be further defined as 'Metabolism of any chemical interacting with an organism which does not occur in the normal metabolic pathways of that organism. The overall effect of the metabolism of xenobiotics is an increase in their water solubility' (Hodgson *et al.*, 1998).

Metabolism of exogenous compounds is required as the routes of absorption and excretion require different chemico-physical properties. As we have seen, absorption across a membrane is favoured by non-polar, lipophilic chemicals; however, excretion routes are water-based and hence favour polar, hydrophilic chemicals. Hence, an important role of metabolism is the conversion of a chemical to alter these properties from those that favour absorption to those that favour excretion. This conversion generally occurs in two stages, and we will consider the role of Phase I (functionalization reactions) and Phase II (conjugation reactions) in chemical toxicity in Chapters 2 and 3, respectively.

1.2.4 Excretion

The major routes of excretion used by the body, urine and faeces, are water based; hence the requirement for metabolism of many chemicals prior to excretion. Small molecules ($Mr < 300$) tend to be excreted via the kidneys, whereas larger molecules are excreted from the liver into the bile and from there incorporated into the faeces. If excretion favours polar, hydrophilic chemicals how do these cross the membranes of, for example, the kidney tubules to enter the urine? The answer is that active transport plays a much more prominent role in excretion than was seen in absorption. Transport proteins with wide substrate specificities exist in all the potential excretion routes, and the role of these is to 'pump out' chemicals from the body. These proteins are found in the liver and kidney, to enhance biliary and renal excretion respectively, but also in the intestine, where they act as the 'first line of defence'. By potentially pumping out chemicals as soon as they enter the body through one of the major absorption routes these transporters can potentially prevent toxic chemicals from ever entering the body. *Table 1.3* lists some of the major transport proteins involved in excretion.

1.3 Summary

Whatever the 'functional' role of a toxic chemical, be it a therapeutic drug with an adverse effect or an environmental pollutant released during industrial processes, the need to understand the way the body deals with this threat is paramount. Not only will this allow us to determine what a 'safe'

Table 1.3   Transport proteins used in excretion

Intestinal efflux		
P-gp	OCT1	OATP3
MRP1	MRP2	
Biliary excretion		
P-gp	MDR3	MRP2
sPGP		
Renal secretion		
OAT1	OAT3	OCT1
OCT2	OATP	P-gp
MRP1		

MDR, multidrug resistance protein; MRP, multidrug resistance related protein; P-gp, P-glycoprotein; OAT, organic anion transporter; OATP, organic anion transport protein; OCT, organic cation transporter. Data from Ayrton and Morgan (2001).

exposure level is, but it may enable us to develop new chemicals that still possess the 'functional' role of the chemical but without the toxic effects. Finally, by understanding adverse processes within the body we may gain important insights into normal physiological functioning, and indeed disease states where these 'normal' processes are disrupted.

We have briefly discussed ADME, the process through which chemicals enter into the body, are processed and finally excreted. In Chapters 2 and 3 we will examine the role of this process in detoxifying chemicals, and also how metabolism can *increase* the toxicity of certain chemicals (toxication or bioactivation reactions). In addition to the role of ADME in detoxifying chemicals, there exists a whole host of cellular systems that can be activated to protect the body from toxic insult, and these will be examined in Chapter 4. Finally, Chapter 5 will examine specific examples of toxicity, showing how different chemicals can cause fundamentally different toxic effects.

In the first five chapters, exposure to toxic chemicals will be presumed to occur one at a time. However, this is rarely the situation, and we are often exposed to a mixture of many chemicals at any one time. In Chapter 6 we will examine how these mixtures of chemicals alter the toxic effects we see, and how we can predict risk and toxicity in such complicated circumstances.

Chapter 7 will examine the role of genetics in controlling and modulating the response to toxic insult. We will examine the basic mechanisms by which gene expression is controlled, and how these expression profiles are altered following toxic insult. In addition, we will examine the variation that exists in DNA between species, and ask how such differences impact upon the extrapolation of risk assessment data from model species. Finally, we will examine how genetic polymorphisms affect the response of individuals to a toxic insult.

This is not a book to be read cover to cover in a single sitting. Instead I encourage the reader to dip into sections that interest them, while using Chapter 8 as a reference to understand the technologies which have made 'molecular' toxicology a reality. Thus, it is hoped, not only will the theory of toxicology, and more specifically its molecular investigation, be understood but the reader will know enough about these 'molecular technologies' to understand their potential application in the investigation of any particular problem.

References

Ayrton, A. and Morgan, P. (2001) Role of transport proteins in drug absorption, distribution and excretion. *Xenobiotica* 31(8/9): 469–497.

Gibson, G.G. and Skett, P. (2001) *Introduction to Drug Metabolism,* 3rd edn. Nelson Thornes, Cheltenham.

Hodgson, E., Mailman, R. and Chambers, E. (1998) *Dictionary of Toxicology*, 2nd edn. Macmillan Reference Ltd., London.

Role of Phase I metabolism in toxicity

2

2.1 What is Phase I metabolism?

As described in Chapter 1 most endogenous and exogenous compounds require chemical alteration before they can be excreted via the hydrophilic urine or faeces. Such metabolism results in increased polarity compared to the parent compound, thus increasing solubility in water and ultimately the rate of excretion. Metabolism occurs in two major phases, functionalization and conjugation. During Phase I metabolism chemically reactive groups are added to, or revealed in, the parent compound, thus producing targets for Phase II conjugation reactions. These reactions encompass oxidation, reduction, hydrolysis, hydration and isomerization.

2.2 Cytochrome P450-mediated Phase I metabolism

2.2.1 Nomenclature and structure of cytochrome P450s

Cytochrome P450s (CYPs) constitute the largest group of enzymes involved in both endobiotic and xenobiotic metabolism. All CYPs are holoenzymes, consisting of a single sub-unit apoprotein and a prosthetic central haem molecule (i.e. they are haemoproteins); indeed, they gain their generic P450 from the absorbance spectral peak at 450 nm caused by the haem Fe^{2+}–CO complex.

Since the discovery of the first CYP enzyme by Mason in 1957 a vast number of CYPs have been identified, from a wide variety of species. Currently over 1900 CYPs have been characterized and named, although the many sequencing projects across the globe means this number is rapidly increasing. Why are there so many CYPs? Two factors contribute to the large number of CYPs, evolutionary conservation and varied substrate exposure.

Firstly, CYPs have been identified from almost all organisms that have been studied, from archaebacteria to humans. Such conservation throughout evolution shows how important these molecules are for the survival of all organisms. This is perhaps not surprising when you revisit the definition of metabolism given in Chapter 1; 'the total chemical transformations of normal body constituents taking place in a living organism, whether they are synthetic (anabolic) or degradative (catabolic) reactions' (Hodgson *et al.*, 1998). CYPs underpin the initial stages in the anabolic and catabolic processing of many compounds within the body, as well as dealing with many foreign compounds that enter the body (xenobiotics).

Secondly, varied substrate exposure underlies the large number of CYPs present within a single organism. As CYPs are responsible for the metabolism of a vast number of different chemical structures then it is logical to presume that a large number of different enzymes would be required within any organism. However, to have a different CYP for every chemical would require many more CYPs than the total estimated protein number in the body. In addition, such a system would not be flexible enough to cope with a new chemical – in this scheme an entire new enzyme would be required. Instead, CYPs have the unusual property of possessing relatively large substrate-binding pockets, allowing some flexibility in the substrate accepted. Through this wide substrate specificity it is therefore possible for CYP-mediated Phase I metabolism (i.e. 100 000s of compounds) to be carried out by a relatively small number of CYPs, and for the system to cope with novel chemicals. In man, 50 genes producing active CYP proteins currently have been identified, plus 15 pseudogenes. Bioinformatic analysis of human EST databases suggest that while a few more CYPs remain to be characterized the majority of human CYPs have now been identified.

Such a large number of related enzymes required a hierarchical nomenclature system to be developed, which is based upon the percentage identity of two CYP sequences at the protein level. CYPs are grouped in the same family if they possess 40% identity and within the same subfamily if they possess 70% identity. To denote this, each CYP is labelled a generic term, family, subfamily, individual: for example, CYP1A1 is a cytochrome (CYP) from family 1 (CYP1), subfamily A (CYP1A) and is the first enzyme in this subfamily (CYP1A1). Such a grouping allows us to show orthologous CYPs, with a CYP3A in rats carrying out nearly identical functions to a CYP3A in humans. A complete list of CYPs is available at the Cytochrome P450 website (http://drnelson.utmem.edu/CytochromeP450.html), hosted by Dr Nelson who devised the nomenclature system. This site is regularly updated and provides the latest information on identification and classification of CYPs. *Table 2.1* shows the major CYPs present in human liver, along with example substrates which are involved in both detoxification and toxication reactions.

This nomenclature system differentiates CYPs purely on the basis of their amino acid sequence and has no direct correlation to enzyme specificities or evolution of the individual families. While it seems likely that CYPs within a single family will have similar properties, due to the high identity of their sequences, how do we relate the properties of one CYP family to another? To address this question a further level of hierarchy has been suggested in the nomenclature of CYPs: namely clans. Many CYPs naturally cluster into clans, as described by their substrate specificity, evolutionary divergence, tissue or sub-cellular localization, etc.; for example, all mitochondrial-expressed CYPs may be clustered together as a single clan. Such higher order clustering helps us to gain insights into the general roles of CYPs, particularly as the number of actual CYP families is rapidly increasing, causing difficulty in comparing inter-, or even intra-, species families. *Table 2.2* shows the major CYP clans identified to date.

While the characterization of CYPs at the genetic level has allowed the identification of consensus regions that presumably play important roles in the functioning of these enzymes it is the combined use of crystallization and *in silico* modelling that has allowed great steps to be made in understanding the biochemical functioning of this superfamily of enzymes.

The crystallization of proteins to determine their structure and thus gain insights into their mode of action has been carried out for many decades.

Table 2.1   Major human hepatic cytochrome P450s

Name	Abundance	Detoxification	Toxication	Accession
CYP1A1	Inducible	Caffeine	Benzo(a)pyrene	NM 000499.2
CYP1A2	15%	Caffeine	2-aminofluorene	NM 000761.2
CYP1B1	Inducible			NM 000104.2
CYP2A6	5%	Coumarin	Nitrosodiethylamine	NM 000746.4
CYP2B6	Inducible	Nicotine	6-aminochrysene	NM 000767.4
CYP2C8				NM 000770.2[b]
CYP2C9	20%[a]	Phenytoin	None known	NM 000771.2
CYP2C18				NM 000772.1
CYP2C19	5%	Mephenytoin		NM 000769.1
CYP2D6	5%	Debrisoquine	NNK	NM 000106.2
CYP2E1	10%	Ethanol	Paracetamol	NM 000773.2
CYP3A4				NM 017460.3
CYP3A5	30%	Erythromycin	Alfatoxin B1b	NM 000777.2
CYP3A7				NM 000765.2

Major hepatic CYPs are listed, along with their relative hepatic abundance, example of substrates metabolized in both detoxification and toxication reactions and RefSeq mRNA accession number.
[a] In the case of CYP2C8/9/18 and CYP4/5/7/43 large inter-individual variation exists in the expression of these enzymes. Hence, an average expression and representative substrate is provided for the entire group rather than individual enzymes.
[b] The CYP2C8 gene codes for multiple transcripts; only the RefSeq for the major transcript being indicated here. NNK = 4-(methylnitrosamino)-1-(3-pyridyl)-1-butanone. Adapted from Gibson and Skett (2001) and Gonzalez and Gelboin (1994).

However, the crystallization of membrane-bound proteins, such as CYPs, has provided the greatest challenge and hence progress in this area has been the slowest. As mentioned above CYPs show a high degree of evolutionary conservation and are seen in a wide variety of species. In the more ancient species, such as the archaebacteria, CYPs exist as non-membrane bound cytosolic forms, and it was here that the initial crystallization experiments focused. In 1987, Poulos and colleagues succeeded in crystallizing a CYP from *Pseudomonas putida*, CYP101 (originally called P450cam as its main substrate was camphor; Poulos *et al.*, 1987). Analysis of this crystal structure, plus primary protein sequence comparisons between this and the mammalian CYPs allowed many important features to be identified within the CYP superfamily. All CYPs are comprised of a series of alpha helices (denoted A–L) and folds, producing an overall gross similarity in structure. Helices I and L are required for contact with the prosthetic haem, while six 'substrate recognition sequences' have been determined throughout the structure by site-directed mutagenesis which are key to substrate binding.

Table 2.2   Cytochrome P450 clans

Clan name	Clan member families
CYP2	CYP1, CYP2, CYP17, CYP18, CYP21
CYP3	CYP3, CYP5, CYP6, CYP9, CYP30
CYP4	CYP4, CYP29, CYP31, CYP32, CYP37
CYP7	CYP7, CYP8
Mitochondrial	CYP10, CYP11, CYP12, CYP24, CYP27, CYP44
C. elegans	CYP14, CYP23, CYP33, CYP34, CYP35, CYP36

While the crystal structure of an archaebacterial, cytosolic, CYP provides important information on the possible structure and function of mammalian CYPs such extrapolation had to be viewed with a degree of caution. CYP101 only shares some 17% identity with mammalian CYPs and the lack of a membrane tether is a major difference. It was therefore possible that CYP101 does not function in exactly the same manner as its counterparts in higher organisms. A large step forward in CYP structural study was made in 2000 when Cosme and Johnson produced the first crystal structure of a mammalian CYP, the rabbit CYP2C5 (Cosme and Johnson, 2000). Remarkably, when the overall structure was compared to that of the archaebacterial CYPs a high degree of similarity was observed, providing strong evidence that models based upon CYP101 are indeed good indicators of mammalian CYP structure.

2.2.2 Cytochrome P450 catalytic cycle

CYPs are capable of catalysing a large variety of reactions (*Table 2.3*), but have a preference for oxidation reactions. These reactions result in the addition of chemically reactive centres for use during Phase II metabolism, and oxidation reactions have the general stochiometry:

$$RH + NADPH + H^+ + O_2 \rightarrow ROH + NADP^+ + H_2O$$

where R = the rest of the substrate's chemical structure.

To carry out these reactions CYPs undergo a complex catalytic cycle, using molecular oxygen and reducing equivalents (supplied by NADPH) to alter the ionic state of the iron atom at the centre of the prosthetic haem. The full catalytic cycle is shown in *Figure 2.1* and can be divided into six phases. Initially, substrate binds to the fully oxidized prosthetic haem, through substrate-induced spin state changes in the ferric haem (step 1). The substrate-bound ferric haem is then reduced by NADPH to ferrous haem (step 2) and molecular oxygen incorporated into the binary ferrous–CYP–substrate complex (step 3). Molecular rearrangement of the complex (step 4), further reduction by addition of a second electron (step 5), and incorporation of the oxygen atom into the substrate (step 6) results in release of the oxygenated metabolite and reformation of the ferric haemoprotein in three poorly understood steps.

2.2.3 Cytochrome P450 pharmacogenetics

As the cytochrome P450 superfamily plays such a fundamental role in the Phase I metabolism of endogenous compounds and xenobiotics then it is logical to hypothesize that genetic variation in the genes encoding for these enzymes may have significant effects on the body's ability to metabolize chemicals. Indeed, polymorphisms have been identified in both the coding

Table 2.3 Cytochrome P450 catalysed reactions

Aromatic hydroxylation	Epoxidation
Aliphatic hydroxylation	Oxidative deamination
O-Dealkylation	N-Oxidation
N-Dealkylation	S-Oxidation
S-Dealkylation	Phoshophionate oxidation
Dehalogenation	Alcohol oxidation

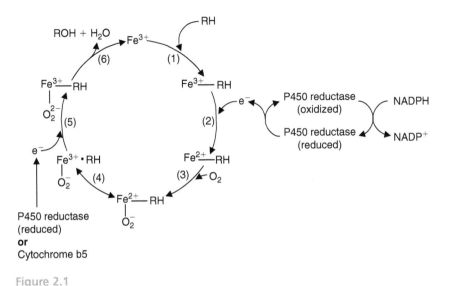

Figure 2.1

Cytochrome P450 catalytic cycle (adapted from Gibson and Skett, 2001).

and regulatory regions of genes encoding all of the CYPs so far examined. An up-to-date list of the identified polymorphisms in human CYPs, along with their functional data and literature references is maintained on the World Wide Web by the Human Cytochrome P450 Allele Nomenclature Committee (http://www.imm.ki.se/CYPalleles/).

The effects of these polymorphisms may be divided into 'silent' (no functional effect), 'enhancing' (increased enzyme amount and/or activity) or 'decreasing' (lowered enzyme amount and/or activity). Of these, the decreasing polymorphisms probably produce numerically the most clinically relevant effects, with several alleles of each CYP showing a decreasing polymorphism.

The net result of such a preponderance of polymorphisms is that individual populations, which may have increased frequencies of certain alleles may show differential responses to chemicals, both in terms of therapeutic effect and toxicity. Therefore, it is important to consider the effects of all non-silent alleles of a CYP when determining the overall response of a population to chemical exposure.

2.2.4 CYP1 family-mediated toxicity

The CYP1 family is perhaps the CYP family most associated with toxic responses. This is because these enzymes are capable of metabolically activating a large number of pro-carcinogens into ultimate carcinogens, which may then form DNA and/or protein adducts. Such chemicals include polycyclic aromatic hydrocarbons (PAHs), polychlorinated biphenyls (PCBs) and, as described below, heterocyclic aromatic amines (HCAs).

HCAs are produced during the cooking of protein-rich foodstuffs (i.e. meat and fish) and have been shown to cause liver, colon, breast and prostate cancer in model animals. Considering the large number of different HCAs, how then do you study which CYPs are responsible for their metabolism, and which of the metabolites produced are carcinogenic? Due to the high complexity to be resolved these questions required the development of *in vitro* systems. By expressing individual human CYPs in bacteria it was possible to determine which CYP produces which metabolite. Concurrently, the genotoxicity of the resultant metabolite(s) was assessed by examining the level of expression of a bacterial gene involved in DNA repair as a marker. In this way, the role of each CYP in the genotoxic effects of a particular HCA could be assessed. Using such a system Oda *et al.* (2001) examined the metabolism of seven HCAs by seven of the major CYPs in human liver. Although all CYPs were capable of metabolizing the HCAs to a certain degree, only CYP1A2-mediated metabolism caused a significant level of genotoxicity. Hence, using a system where individual components of the total metabolism of a compound are isolated it is possible to determine which component is responsible for toxication pathways, and by inference which are detoxifying.

The corollary of the above methodology, the study of a single enzyme in isolation, is the study of the effects on a total system of the removal of a single enzyme. Throughout this book we will examine the use of transgenic animals to examine the role of individual enzymes in toxicity. The ability to completely remove a single protein from an organism is an extremely powerful tool and allows researchers to exactly define the role this protein has in both normal physiology and upon toxic insult.

A prime example of the power of transgenic technology is provided by the study of the arylamine 4-aminobiphenyl (ABP)-induced toxicity. ABP is formed during several industrial chemical processes such as rubber manufacture and workers with high occupational exposure to ABP and other aromatic amines have an increased risk of urinary bladder cancer (up to 90-fold). In addition, ABP is present in tobacco smoke and smokers have up to a 7-fold increased risk of both ABP-DNA adducts in the urinary bladder and urinary bladder cancer. Thus, the overall exposure of the population to ABP may be quite high and its mode of toxicity needs to be fully understood to allow a clear risk assessment to be made.

Hepatic metabolism of ABP consists of both toxication and detoxification pathways (*Figure 2.2*). ABP toxication pathways require metabolic activation via *N*-hydroxylation, followed by *O*-acetylation or *O*-sulphation thereby forming highly reactive species. *In vitro* experiments with purified human CYPs had implicated CYP1A1, CYP1A2 and CYP1B1 in the initial *N*-hydroxylation step and as in human liver CYP1A1 and CYP1B1 are present only at very low levels CYP1A2 was identified as the CYP most likely to be responsible for this metabolic activation. Kimura *et al.* (1999) and Shertzer *et al.* (2002) used transgenic technology to test this hypothesis; *Cyp1a2(−/−)* (knock-out) and *Cyp1a2(+/+)* (wild-type) mice were exposed to ABP and markers of toxicity measured. Markers used included number of carcinoma, adenomas and preneoplastic foci (Kimura *et al.*, 1999) and hepatic thiol, serum ornithine carbamoyltransferase and methaemoglobin formation (Shertzer *et al.*, 2002). No major differences were observed between wild-type and knock-out animals, strongly suggesting that CYP1A2 *does not* play a major role in the toxication pathway of ABP *in vivo*. Such apparent contradiction may be as a result of the differing test systems (*in vitro* human CYPs versus *in vivo* murine knockouts) or may suggest that either an alternate CYP (CYP1A1 or CYP1B1?) or flavin mono-oxygenase is responsible for the toxication of ABP; further research can now be focused to reconciling these differences.

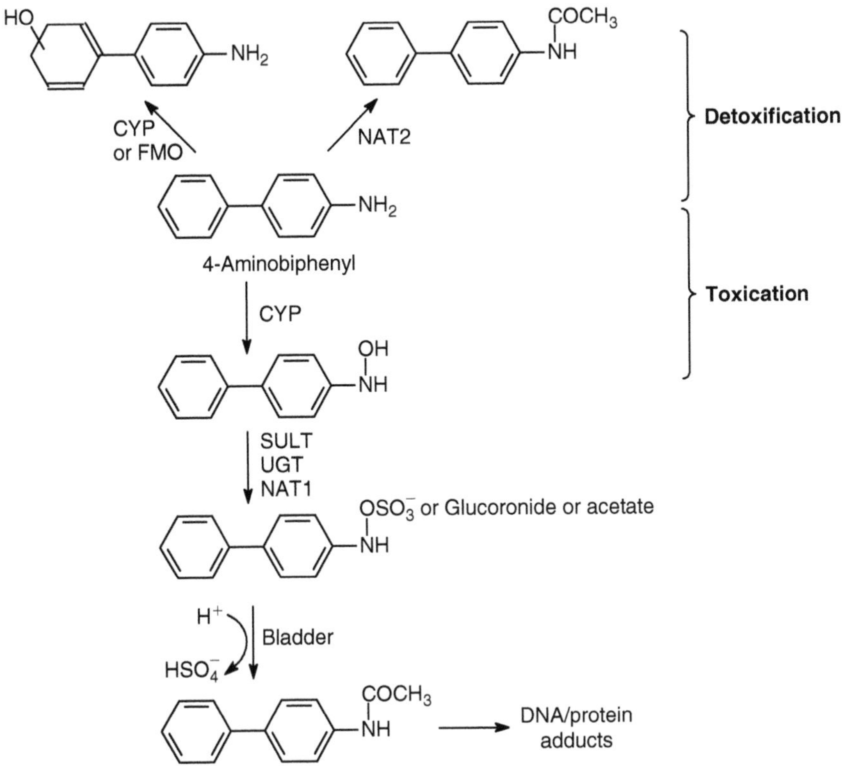

Figure 2.2

Hepatic metabolism of 4-aminobiphenyl

2.2.5 CYP2 family-mediated toxicity

As with the CYP1 family, CYP2 family members are responsible for the metabolic activation of a number of compounds, resulting in toxic metabolites. Perhaps the best-known example of a CYP2 family member-mediated metabolic activation of a compound resulting in toxicity is the CYP2E1-mediated bioactivation of paracetamol. Paracetamol is a widely used analgesic/antipyretic and under normal usage is generally considered safe. However, high exposure to paracetamol may lead to massive hepatic necrosis, which is often fatal. As the toxic effects only occur at high doses this suggests that under normal conditions the toxic metabolites are not produced in high enough concentrations, or are removed too quickly to cause overt toxicity. To establish the molecular mechanism of paracetamol toxicity it was therefore necessary to study not only the enzymes that metabolize paracetamol and its metabolites, but also their relative abundances. *Figure 2.3* shows the major routes of metabolism of paracetamol in the liver: it can be seen that CYP2E1 is responsible for the formation of the reactive intermediate N-acetyl-p-benzoquinoneimine (NAPQI). Under normal physiological conditions this metabolite is quickly removed by Phase II conjugation to glutathione. However, in situations where glutathione levels are depleted, excess NAPQI accumulates and protein adducts form, ultimately resulting in hepatic necrosis.

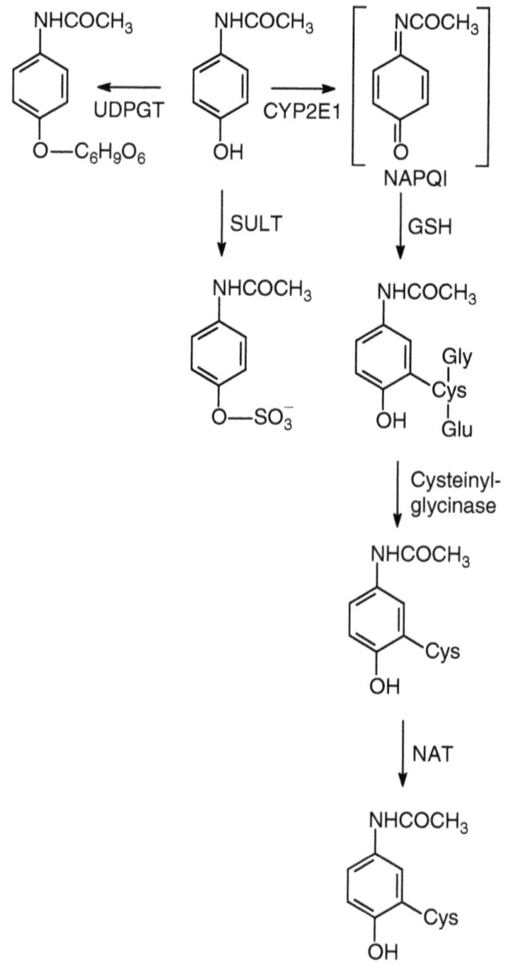

Figure 2.3

Hepatic metabolism of paracetamol.

Cyp2e1 (−/−) transgenic knock-out mice exposed to paracetamol were more resistant to hepatic necrosis than wild-type mice, showing the important role of CYP2E1 in this toxicity (Zaher *et al.*, 1998). A surprising result in this study, however, was that *Cyp2e1* (−/−)/*Cyp1a2* (−/−) double knock-out mice demonstrated even higher resistance to paracetamol-induced hepatic toxicity. This finding implicated CYP1A2 in paracetamol-induced hepatic toxicity for the first time. Hence, the use of transgenic animals can once again be seen to reveal hitherto unsuspected complexities in toxication pathways.

2.2.6 CYP3A subfamily-mediated toxicity

In general, CYP3A metabolism is associated with detoxification reactions and not the metabolic activation of compounds to toxins. Indeed, toxicity associated with CYP3A is usually dependent upon exposure to multiple

CYP3A substrates simultaneously, resulting in drug–drug interactions: this topic will be covered fully in Chapter 6. However, due to the large substrate profile of CYP3A enzymes it is perhaps not surprising that some examples of metabolic activation leading to toxicity exist.

Aflatoxin B1 (AFB) is a fungal metabolite produced by strains of *Aspergillus flavus*. As these fungi commonly contaminate corn and peanut crops exposure to humans is a distinct possibility. Coupled to this, the knowledge that AFB is one of the most potent rodent hepatocarcinogens known necessitates the need for accurate risk assessment in man. AFB may be biotransformed into a number of metabolites in man, although only the AFB-8,9-epoxide has carcinogenic potential (*Figure 2.4*). It is therefore important to determine which CYPs are responsible for the production of this metabolite. Using the bacterial-expression system previously described to study CYP1A2-mediated toxicity of heterocyclic amines, the CYPs responsible for metabolic activation of aflatoxin B1 were examined (Oda *et al.*, 2001). Using this system, CYP1A2 and CYP3A4 were both shown to be capable of activating aflatoxin to the AFB-8,9-epoxide, resulting in genotoxicity. However, the abundances of CYP3A4 and CYP1A2 in human liver are markedly different, with CYP3A4 being 2–4 times more prevalent (*Table 2.1*). Hence, *in vivo* it is more likely that CYP3A4 carries out the majority of metabolic activation of aflatoxin B1.

As CYP3A4 is the most abundant CYP in human liver, how then can we protect ourselves against this potent hepatocarcinogen? The antiparasitic drug Oltipraz has been shown to confer limited resistance to AFB exposure in rats. The molecular mechanism behind this has not been fully elucidated, but

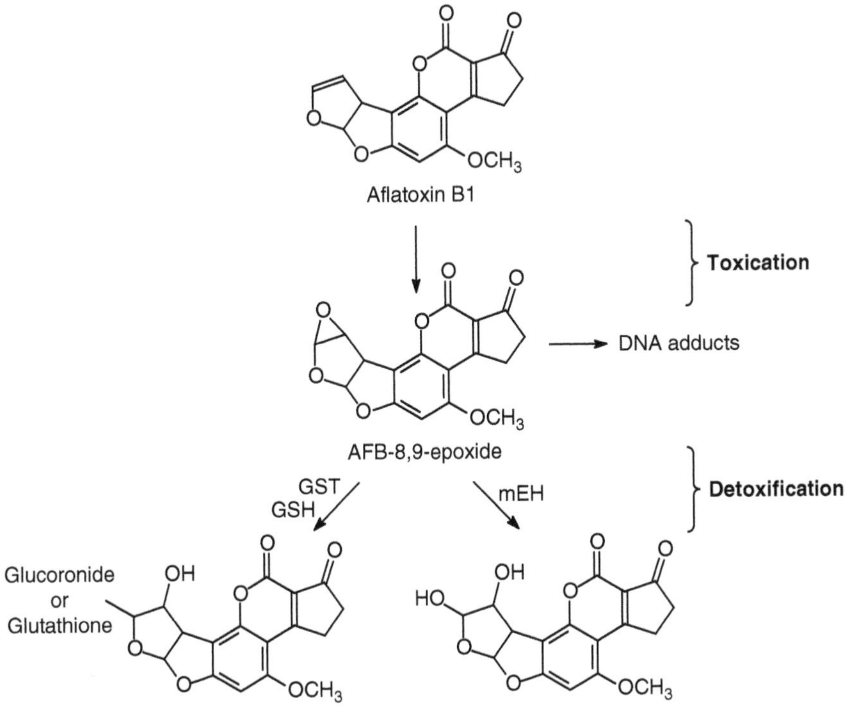

Figure 2.4

Hepatic metabolism of aflatoxin B1.

appears to involve the ability of Oltipraz to induce expression of both microsomal epoxide hydrolase and μ-class glutathione S-transferases. As can be seen from *Figure 2.4*, these enzymes are responsible for the Phase II metabolism of the AFB-9,8-epoxide, resulting in detoxification. Hence, by altering the balance of drug-metabolizing enzymes within the liver it may be possible to reduce or negate the toxic effects of a compound, a paradigm that will be frequently revisited in this book.

2.3 Flavin mono-oxygenase-mediated Phase I metabolism

Phase I metabolism is often thought of as the sole domain of cytochrome P450 enzymes. However the purification of the first flavin mono-oxygenase in 1969 provided an alternative enzyme family for catalysing the oxidation of heteroatom-containing chemicals.

2.3.1 Nomenclature and structure of Flavin mono-oxygenases

Unlike CYPs there are relatively few FMO enzymes; humans express only five FMO enzymes, and only three of these in adult liver (*Table 2.4*) compared to the 50+ CYP enzymes. Due to this there exists a much-simplified nomenclature, with enzymes termed FMOx, where x is a number (currently 1–5). FMOs with less than 50% identity at the amino acid sequence are segregated as new families, while orthologues require greater than 80% identity. Using this nomenclature, orthologues will be given the same name, which can lead to some confusion as FMO1 may refer to the rabbit, pig or human enzyme.

FMOs are highly lipophilic, membrane-bound proteins which, like CYPs, have proved difficult to crystallize and hence determine their 3-dimensional structure. In addition, no counterpart of CYP101 is known, an ancestral, cytosolic FMO which could be easily crystallized and structural information extrapolated to higher animals. To circumvent these problems, structural information had to be extrapolated by comparison to non-FMO flavoproteins which had been crystallized, such as *E. coli* glutathione reductase (Mittl and Schulz, 1994), although the accuracy of such models is difficult to judge. However, based upon such information, and the sequencing of mammalian FMOs, putative FAD and NADP+ binding domains have been identified. In addition, it has been shown that, in contrast to CYPs, the N-terminal hydrophobic region of the molecule is not responsible for membrane insertion.

Table 2.4   Human flavin mono-oxygenases

Name	Distribution	RefSeq accession
FMO1	Fetal liver	NM 002021.1
FMO2	Muscle, kidney, lung	NM 001460.1
FMO3	Liver, kidney, brain	NM 006894.2
FMO4	Liver, kidney, prostate, brain	NM 002022.1
FMO5	Liver, kidney, spleen, brain	NM 001461.1

Human FMOs are listed, along with their distribution and RefSeq mRNA accession number.

2.3.2 Flavin mono-oxygenase catalytic cycle

The FMO catalytic cycle is similar to that of the CYP enzymes, with an overall stochiometry of:

$$RH + NADPH + H^+ + O_2 \rightarrow ROH + NADP^+ + H_2O$$

where R = the rest of the substrate's chemical structure.

One difference between, CYP and FMO-catalysed reactions is that FMO-catalysed reactions have an increased specificity towards N- and S- as the reactive centre. Beyond this however, there are two major differences between the two catalytic cycles. Firstly, the prosthetic core in FMOs is a flavin molecule as opposed to the haem seen in CYPs. However, both moieties act as electron acceptors from NADPH, as well as interacting with molecular oxygen and the substrate. Secondly, while the net effect of the two catalytic cycles is the same the order of reactions within the cycles is different. *Figure 2.5* shows the full FMO catalytic cycle, which may be broken into four separate reactions. Initially, fully oxidized flavoprotein is reduced using NADPH (step 1), and then addition of molecular oxygen produces a hydroperoxyflavin intermediate (step 2). This hydroperoxyflavin intermediate is open to neutrophilic attack by substrates at the terminal oxygen atom, and an oxygenated metabolite can be released (step 3). Finally, water is released and the fully oxidized flavoproteins can be reformed (step 4). It can thus be seen that in the FMO catalytic cycle the initial reduction of the flavin moiety occurs prior to substrate binding, whereas these steps are reversed in the CYP catalytic cycle.

2.3.3 Flavin mono-oxygenase pharmacogenetics

In comparison to the cytochrome P450s much less work has been undertaken to characterize genetic polymorphisms in the genes encoding the flavin mono-oxygenase enzymes. However, functional polymorphisms have been identified within these genes, and in particular the *FMO3* gene, the major human flavin mono-oxygenase, as this results in the disorder known as 'fish odour syndrome'. FMO3 is responsible for the metabolism of trimethylamine (TMA), resulting in the excretion of trimethylamine N-oxide in the urine. However, in some individuals this conversion does not occur, with up to 90% of TMA being excreted unchanged. While TMA-N-oxide is an odourless

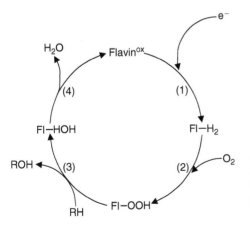

Figure 2.5

Flavin mono-oxygenase catalytic cycle (adapted from Ziegler, 1990).

excretion product TMA itself has a potent odour, and individuals excreting large amounts of TMA also exude an unpleasant odour; hence the name of the disorder (fish odour syndrome). Al-Waiz *et al.* (1987) identified that this error in metabolism was due to a polymorphism within the *FMO3* gene, and further work by many researchers has identified a range of polymorphisms that result in an inability to metabolize TMA correctly, resulting in fish odour syndrome (*Table 2.5*).

Table 2.5 FMO3 polymorphisms resulting in 'fish odour syndrome'

Region of gene	Substitution	Region of gene	Substitution
Exons 1 + 2	Deletion	Exon 7	E305X
Exon 3	A52T	Exon 7	E314X
Exon 3	N61S	Exon 7	R387L
Exon 3	M66I	Exon 9	M434I
Exon 3	M82T	Exon 9	R492W
Exon 4	P153L		

FMO3 polymorphisms are listed by location within the gene, and the resultant amino acid substitution indicated. Data from Cashman (2002).

2.3.4 Flavin mono-oxygenase-mediated toxicity

It is true to say that fewer examples of FMO-mediated toxicity exist than CYP-mediated ones and in general FMO-mediated metabolism is associated with detoxification pathways as opposed to toxication. Indeed, it has been hypothesized that FMO evolution may have been driven to produce enzymes that safely metabolize substrates that would otherwise lead to toxic or auto-inhibitory metabolites if metabolized by CYPs (Ziegler, 1990). However, some examples of substrates where FMO-mediated metabolism may result in toxicity do exist, including the FMO-mediated metabolism of thione-containing compounds.

Thiocarbamates are found in many therapeutic drugs, agrichemical and industrial chemicals. Thiourea is the parent structure for many of these thiocarbamates and it has been demonstrated to cause thyroid and liver tumours in rats. In addition, metabolites of thiourea have been shown to cause pulmonary oedema. Due to the high potential for human exposure to these compounds there exists a need to understand the molecular mechanisms of these toxic responses in order to correctly formulate risk assessments for human exposure.

Isolated enzyme studies implicated FMOs in the initial *S*-oxidation of the thionocarbonyl group of thiourea. To further study this reaction Smith and Crespi (2002) engineered mouse embryonic C3H10T1/2 cells which expressed human FMO3 and their response to thiourea exposure was examined. Cells expressing high levels of human FMO3 exhibited increased susceptibility to thiourea-induced toxicity as assessed by clonogenic assay, strongly suggesting that FMO3 was critical in thiourea-mediated toxicity. Experiments with cells expressing lower levels of FMO3 produced a further insight into the mechanism of toxicity; toxicity was only observed when glutathione was depleted. Thus, as was previously seen with paracetamol toxicity, production of a toxic metabolite is not the only factor for determining if a toxic response is seen; the rate of its removal is also important. In situations where glutathione is depleted, toxic metabolite accumulation occurs, resulting in a toxic response. The proposed metabolism pathway of thiourea is shown in *Figure 2.6*.

2.4 Other Phase I-mediated toxicity

While the majority of Phase I metabolism, and hence Phase I metabolic activation of compounds to toxins, are oxidation reactions mediated by either CYPs or FMOs there are a number of other enzymes that play roles in these reactions and these too are responsible for their own detoxification and toxication reactions.

2.4.1 P450 reductase-mediated toxicity

Cytochrome P450 reductase is an important part of the catalytic cycle for CYP functioning: it transfers reducing equivalents from NADPH to the haemoprotein (*Figure 2.1*). However, CYP P450 reductase can also directly reduce compounds, and hence may play a role in both detoxification and toxication pathways. One area where P450 reductase appears to play an important role in toxicity is in the brain.

The brain has a number of important neurotransmitters, one of which is dopamine. These chemicals transmit signals between neurons and thus process information within the brain. Dopamine-mediated neurotransmission is of considerable interest as it seems to underlie the mental disorder schizophrenia. Whilst the molecular mechanisms causing schizophrenia are not fully understood, and indeed it appears to be a multifactorial disease, drugs that cause schizophrenia-like symptoms act upon the dopaminergic system, suggesting dopamine is a key transmitter in this disease. In the brain dopamine is metabolized to dopamine quinine by prostaglandin H synthetase, which then undergoes a spontaneous ring-formation to produce aminochrome. Aminochrome is metabolized by P450 reductase in a

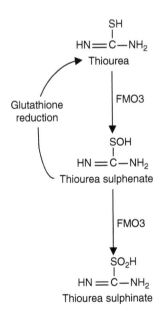

Figure 2.6

Hepatic metabolism of thiourea (adapted from Smith and Crespi, 2002).

single electron reduction producing dopamine-o-semiquinone, and by DT-diaphorase in a two electron reduction producing dopamine-o-hydroquinone (*Figure 2.7*). Both semiquinone and hydroquinone are unstable and can auto-oxidize to reform aminochrome, converting molecular oxygen to superoxide radicals in the process. However, *in vivo*, superoxide dismutase prevents hydroquinone auto-oxidation and this molecule is instead detoxified by conjugation to sulphur: hence, only the semiquinone produces the highly reactive superoxide radicals which cause toxicity (*Figure 2.7*; Baez *et al.*, 1995).

2.4.2 Cytochrome b5-mediated toxicity

As with the previous example, P450 reductase, cytochrome b5 may act as a carrier of reducing equivalents for the CYP catalytic cycle. It too can donate the electron independently, participating in both detoxification and toxication reduction reactions.

Chromium is a dietary component required for the metabolism of carbohydrates and lipids. Chromium is environmentally available in two oxidation states, Cr(III) and Cr(VI). Compounds containing the latter are well absorbed by the body, whereupon they are generally reduced to Cr(III). During this reduction, the reactive Cr(V) and Cr(IV) intermediates are formed, and the resultant release in reactive oxygen species has been linked

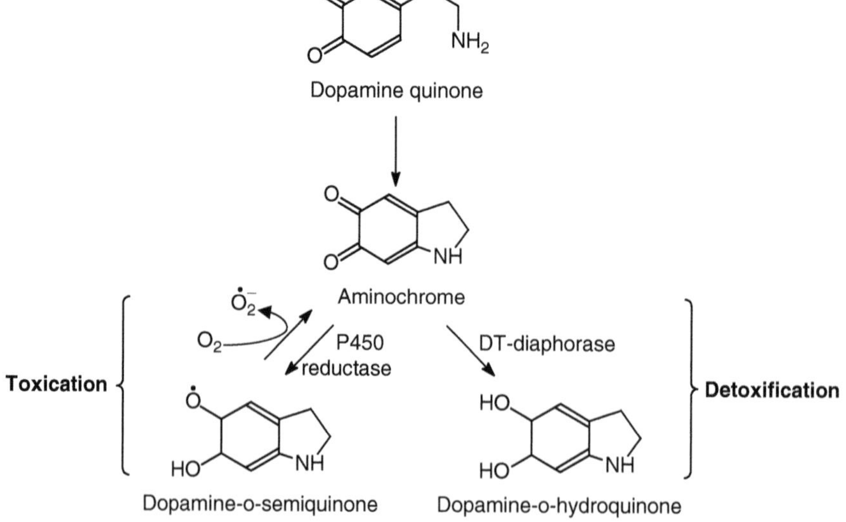

Figure 2.7

Metabolism of dopamine.

to the cytotoxicity, mutagenicity and carcinogenicity of chromium-containing compounds. What then are the molecular mechanisms by which Cr(VI) is reduced to Cr(III)?

Through inhibitor studies P450 reductase was implicated in this reduction reaction; however, purified P450 reductase is a poor reducer of Cr(VI). To examine this apparent contradiction, Jannetto *et al.* (2001) used human liver microsomes to examine the reduction of Cr(VI) to Cr(III). Collection of sub-cellular fractions demonstrated that 88% of chromium reduction occurred in fractions that contained both P450 reductase and cytochrome b5. While this suggests a role for both of these proteins in chromium reduction, it does not rule out alternative pathways. To examine this Jannetto *et al.* then used pro-teoliposomes containing either P450 reductase, cytochrome b5 or both and proved that it was an interaction of these two enzymes that was responsible for the reduction, and hence the production of the reactive intermediates. It was therefore possible to postulate the transfer of reducing equivalents seen in *Figure 2.8*.

2.4.3 Prostaglandin synthetase-mediated toxicity

Prostaglandin H synthase is, like CYPs, a haemoprotein and is a key enzyme in the biosynthesis of prostanoids (Degen *et al.*, 2002). Prostanoids (pros-taglandins and their related compounds) are important modulators in many physiological processes through their ability to modulate the activity of other hormones, and are also involved in the contraction of uterine smooth muscle during labour and inflammatory reactions. Prostaglandin H synthase has two distinct chemical activities, a cyclo-oxygenase and a hydroperoxidase function, that convert arachidonic acid first to pros-taglandin G_2 and then to prostaglandin H_2 (*Figure 2.9*). During this latter step, a co-oxidation reaction may occur, leading to the metabolic activation of certain drugs, such as aminopyrene. In addition to this co-oxidation reaction, metabolic activation of certain compounds, such as paracetamol, may occur via a radical-mediated route (*Figure 2.9*).

2.5 Summary

In this chapter we have examined the initial reaction of the body to xeno-biotics; the activation of systems that are designed to remove the chemical

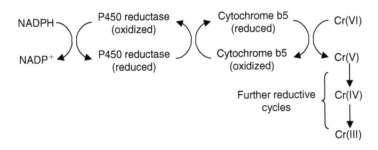

Figure 2.8

Reduction of chromium by cytochrome b5 (adapted from Jannetto *et al.*, 2001).

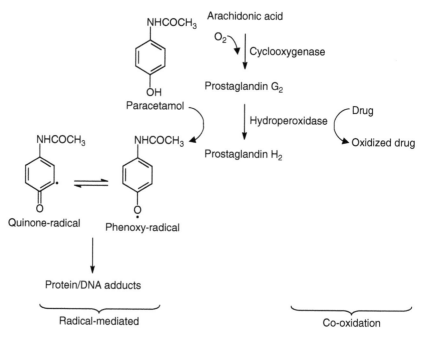

Figure 2.9

Prostaglandin H synthetase-mediated activation of paracetamol (adapted from Degen *et al.*, 2002).

from the body. While these systems may have originally developed to handle the metabolism, both catabolic and anabolic, of endogenous compounds they have since diversified to be able to handle the large number of exogenous chemicals that an individual may be exposed to during their life span. As we have seen, in the majority of cases these systems are adept at handling such compounds and facilitating their removal from the body by increasing their overall polarity. However, a number of chemicals are actually made more toxic (bioactivation or toxication pathways) rather than safer (detoxification pathways) by this metabolism, and it is these reactions that are potentially dangerous to the body.

In the next chapter we will consider how another group of enzymes carry out Phase II metabolism and increase the rate of excretion even further. In addition, as we will see, they play an important role in the detoxification of any compounds that were either not detoxified during Phase I metabolism, or were in fact made more toxic.

References

Al-Waiz, M., Ayesh, R., Mitchell, S.C., *et al.* (1987) A genetic polymorphism of the N-oxidation of trimethylamine in humans. *Clin. Pharmacol. Therapeutics* **42**: 588–594.

Baez, S., Linderson, Y. and Segura-Aguilar, J. (1995) Superoxide dismutase and catalase enhance autooxidation during one-electron reduction of aminochrome by NADPH-cytochrome P-450 reductase *Biochem. Molec. Med.* **54**(1): 12–18.

Cashman, J.R. (2002) Flavin mono-oxygenases. In: Ioannides, C. (ed.) *Enzyme Systems that Metabolise Drugs and Other Xenobiotics.* John Wiley & Sons, Chichester, pp. 67–93.

Cosme, J. and Johnson, E.F. (2000) Engineering microsomal cytochrome P450 2C5 to a soluble, monomeric enzyme. Mutations that alter aggregation, phospholipid dependence and membrane binding. *J. Biol. Chem.* **275**: 2545–2553.

Degen, G.H., Vogel, C. and Abel, J. (2002) Prostaglandin synthases. In: Ioannides, C. (ed.) *Enzymes Systems that Metabolise Drugs and Other Xenobiotics.* John Wiley & Sons, Chichester, pp. 189–229.

Gibson, G.G. and Skett, P. (2001) *Introduction to Drug Metabolism,* 3rd edn. Nelson Thornes, Cheltenham.

Gonzalez, F.J. and Gelboin, H.V. (1994) Role of human cytochromes P450 in the metabolic activation of chemical carcinogens and toxins. *Drug Metab. Rev.* **26**(1–2): 165–183.

Hodgson, E., Mailman, R. and Chambers, E. (1998) *Dictionary of Toxicology,* 2nd edn. Macmillan Reference Ltd., London.

Jannetto, P.J., Antholine, W.E. and Myers, C.R. (2001) Cytochrome b5 plays a key role in human microsomal chromium(VI) reduction. *Toxicology* **159**: 119–133.

Kimura, S., Kawabe, M., Ward, J.M., *et al.* (1999) CYP1A2 is not the primary enzymes responsible for 4-amino-b-phenyl-induced hepatocarcinogenesis in mice. *Carcinogenesis* **20**: 1825–1830.

Mason, H.S. (1957) Mechanisms of oxygen metabolism. *Adv. Enzymol.* **19**: 74–233.

Mittl, P.R. and Schulz, G.E. (1994) Structure of glutathione reductase from *Escherichia coli* at 1.86 A resolution: comparison with the enzyme from human erythrocytes. *Prot. Sci.* **3**(5): 799–809.

Oda, Y., Aryal, P., Terashita, T., *et al.* (2001) Metabolic activation of heterocyclic amines and other precarcinogens in *Salmonella typhimurium umu* tester strains expressing cytochrome P4501A1, 1A2, 1B1, 2C9, 2D6, 2E1 and 3A4 and human NADPH-P450 reductase and bacterial O-acetyltransferase. *Mutation Res.* **492**: 81–90.

Poulos, T.L., Finzel, B.C. and Howard, A.J. (1987) High-resolution crystal structure of cytochrome P450cam. *J. Mol. Biol.* **195**: 687–700.

Shertzer, H.G., Dalton, T.P., Talaska, G., *et al.* (2002) Decrease in 4-aminobiphenyl-induced methemoglobinemia in Cyp1a2(−/−) knockout mice. *Toxicol. Appl. Pharmacol.* **181**(1): 32–37.

Smith, P.B. and Crespi, C. (2002) Thiourea toxicity in mouse C3H/10T12 cells expressing human flavin-dependent mono-oxygenase 3. *Biochem. Pharmacol.* **63**(11): 1941–1948.

Zaher, H., Buters, J.T., Ward, J.M., *et al.* (1998) Protection against acetaminophen toxicity in CYP1A2 and CYP2E1 double-null mice. *Toxicol. Appl. Pharmacol.* **152**: 193–199.

Ziegler, D.M. (1990) Flavin-containing mono-oxygenases: enzymes adapted for multisubstrate specificity. *Ann. Rev. Pharmacol. Toxicol.* **33**: 321–324.

Role of Phase II metabolism in toxicity

3

3.1   What is Phase II metabolism?

In the previous chapter we have discussed the enzymes involved in the initial steps in metabolism of chemicals, the Phase I enzymes whose role is the addition of chemically reactive groups to molecules. While this process increases the overall polarity of the molecule to some degree it is usually not sufficient to cause a large increase in the rate of excretion of the primary metabolite. For this to be achieved large polar groups also need to be added, and Phase II enzymes achieve this through interaction with the chemically reactive groups added during Phase I metabolism.

The pressures placed upon Phase II metabolism by the body are similar to those exerted on Phase I metabolism – the need to handle a large number of structurally different chemicals and to be able to respond to increases in the concentrations of these chemicals. It is therefore perhaps not surprising that the overall nature of Phase II enzymes is similar to that of Phase I enzymes; they are present in most tissues but concentrated in the liver, can accept relatively large substrate profiles and are inducible by their substrates. However, unlike Phase I metabolism, where the majority of reactions are catalysed by enzymes from just one or two families (CYP and FMO), Phase II metabolism is carried out by several smaller families, each of which catalyses the addition of a different conjugate onto the substrate (*Table 3.1*).

While there are several enzyme families that catalyse Phase II metabolism the majority of reactions in the body are catalysed by one of three enzyme families, resulting in the addition of glucuronide, sulphate or glutathione conjugates; these families will be discussed in detail herein. These three reactions also highlight another important difference between Phase I and Phase II metabolism. In Phase I metabolism specificity for an enzyme is dependent upon the overall structure of the chemical. However, a major

Table 3.1   Enzyme families carrying out Phase II metabolism

Reaction	Enzyme	Conjugate
Acetylation	Acetyltransferase	$-OCCH_3$
Amino acid conjugation	Various	Various
Glucuronidation	UDP-Glucuronosyltransferase	$-\beta$-glucuronide
Glutathione conjugation	Glutathione-*S*-transferase	Glu—Cys—Gly
Glycosylation	UDP-Glycosyltransferase	Various
Methylation	Methyltransferase	$-CH_3$
Sulphation	Sulphotransferase	$-SO_3$

determinant in Phase II metabolism is the chemical reactivity of the substrate. Phase II reactions are therefore subdivided into those in which the conjugate is reactive (Type I) and those in which the substrate is reactive (Type II). Unreactive or poorly reactive chemicals may be substrates for almost all of the enzyme families listed in *Table 3.1*, with the exception of the glutathione transferases which are targeted towards the metabolism of chemically reactive substrates. Phase II metabolism is therefore able to deal with all products of Phase I metabolism, including potentially harmful reactive intermediates. Indeed as we will see later in this chapter glutathione conjugation plays a vital role in the body's defence systems against reactive chemicals which might otherwise cause toxicity.

Another major difference between Phase I and Phase II reactions is that whereas Phase I reactions involve the addition of single atoms, Phase II reactions result in the addition of more complex chemical structures (the conjugate). Before this can occur, the conjugate first needs to be synthesized and, with the exception of glutathione, made chemically reactive. As we shall see, conjugate availability may be a limiting step in some Phase II pathways (e.g. sulphation). This is an important factor because, as we saw in Chapter 2, toxicity may occur if the production of reactive intermediates by Phase I exceeds their removal by Phase II. Such a situation can occur if the rate of Phase II metabolism is limited by the ability to synthesize conjugate(s), and may explain why many chemicals are able to undergo Phase II metabolism via several pathways, thus ensuring the rapid and efficient removal (via metabolism and excretion) of potentially toxic chemicals.

3.2   Glucuronide conjugation

Phase II metabolism via glucuronidation represents the largest subset of Phase II reactions. Such widespread usage is driven by the extensive availability of the conjugate (see section 3.2.2 for further details) which means that depletion of conjugate is never a limiting factor in the rate of these particular Phase II reactions.

3.2.1   Nomenclature and structure of UDP-glucuronosyltransferases

As we have already seen with Phase I reactions, in order to carry out the metabolism of the millions of chemicals, both endogenous and xenobiotic, that the body is exposed to during its lifetime requires families of enzymes that possess large, overlapping substrate profiles. This paradigm is also true for Phase II metabolism and at present 16 UDP-glucuronosyltransferase enzymes (UGTs) have been identified in man. Using the same basic nomenclature system previously described for the cytochrome P450 enzymes, UGT enzymes are clustered into families and subfamilies according to their identity at the amino acid level. Enzymes showing >45% identity at the amino acid level belong to the same family, while >60% corresponds to enzymes from within the same subfamily. UGTs are thus named by the nomenclature UGT, family, subfamily, individual, and the currently identified human UGTs are shown in *Table 3.2*.

Table 3.2 Human UDP-glucuronosyltransferase enzymes

Name	Example substrate	Accession
UGT Family 1		
UGT1A1	Ethinyloestradiol	NM 000463.1
UGT1A3	Cyproheptadine	NM 019093.1
UGT1A4	Amitriptyline	NM 007120.1
UGT1A5	Pregnenalone-16α-carbonitrile	M 84129.1*
UGT1A6	Paracetamol	NM 001072.1
UGT1A7	SN-38	U39570.1*
UGT1A8	Oestrone	NM 019076.2
UGT1A9	Paracetamol	NM 021027.1
UGT1A10	Mycophenolic acid	U 39550.2*
UGT Family 2		
UGT2A1	3-hydroxybiphenyl	NM 006798.1
UGT2B4	Androsterone	NM 021139.1
UGT2B7	Morphine	NM 001074.1
UGT2B10	None known	NM 001075.1
UGT2B11	4-hydroxyoestrone	NM 001073.1
UGT2B15	Androsterone	NM 001076.1
UGT2B17	Testosterone	NM 001077.1

Major human UGTs are listed, along with examples of substrates and RefSeq mRNA accession number. Accession numbers marked '*' represent only the variant exon 1, as no RefSeq is available. Adapted from Bock (2001).

Experiments carried out using recombinant UGTs have allowed a greater understanding of the physical structure of this enzyme within the cell. Nascent UGT protein is produced as a precursor including an amino terminal signal peptide which mediates the integration of UGTs into the endoplasmic reticulum (ER). An interesting consequence of this integration is that modelling predicts that the active site would be presented on the luminal side of the ER. As the conjugate, UDP-glucuronide, is synthesized in the cytosol such an orientation would require transporters to transfer the conjugate across the ER membrane. Recent evidence has also suggested that UGTs may integrate as homodimers into the endoplasmic reticulum, dimerization occurring via interactions at the amino termini of the UGTs, although this does not appear to affect the overall characteristics of the enzyme (Meech and Mackenzie, 1997).

One unusual feature of UGT enzymes is the gene locus encoding the UGT1A subfamily. Gene structure and control will be discussed in more detail in Chapter 7, however, the simplest definition of a gene is an enhancer and promoter controlling the expression of a coding region, producing a single mRNA molecule which is translated to produce a single protein. In mammals the vast majority of coding regions of genes are further sub-divided into exons and introns: the former containing the information coding for the protein whereas the latter are non-coding spacers. Following transcription the introns are removed and the exons joined together by splicing to produce the mature mRNA. In a subset of genes an additional level of protein diversity is generated through the use of alternate splicing, whereby individual exons may be excluded or several mutually exclusive versions of a single exon may exist. The consequence of alternate splicing is the generation of a family of proteins which are identical across most of their sequence, but differ in the region encoded by the alternatively spliced exon(s); thus several closely related proteins can be encoded by a single gene locus.

In the case of the UGT1A protein subfamily, we know that these proteins must be closely related as this is the criterion for inclusion within the same sub-family. However, closer examination of the UGT1A protein sequences reveals that all of the variation lies within the N-terminus of the protein, specifically that region encoded for by exon 1 of the gene. All UGT1A enzymes are encoded for by a single gene locus, some 160 kbp in length, as shown in *Figure 3.3* (Ritter *et al.*, 1992). While each mRNA, and thus protein, contains the same information encoded by the constitutive exons 2–5 of the gene, the first exon is variable and provides the unique characteristics of that UGT1A family member. Transcription is initiated from a promoter next to one of the alternate first exons and all other redundant exon 1 sequences are removed by splicing of the pre-mRNA to form a unique mRNA. This provides an excellent example of protein diversity encoded by minimal DNA variation (see *Figure 3.3*; Tukey and Strassburg, 2000).

3.2.2 UDP-glucuronosyltransferase-catalysed reactions

Glucuronidation reactions involve the transfer of glucuronide from UDP-glucuronide to the substrate at —OH, —COOH, —SH and —NH$_2$ groups, with the overall stoichiometry

$$R—OH + UDPGA \rightarrow R—OGA + UDP$$

where R = the rest of the substrate's chemical structure and UDPGA the conjugate UDP-glucuronic acid.

As described in the introduction, the first step in all Phase II reactions is the formation, and activation if necessary, of the conjugate. For glucuronidation the conjugate, glucuronide, is formed from glucose-1-phosphate. As all cells require glucose-1-phosphate for use in glycolysis, high levels of glucose-1-phosphate are present in all cells. Such a ubiquitous expression means that conjugate supply is never a limiting factor for UGT-catalysed reactions. *Figure 3.1* shows the pathway by which glucose-1-phosphate is converted to UDP-glucose and finally the α-UDP-glucuronic acid.

Glucuronidation is thought to occur via an acid–base reaction. In this, a basic group within the active site of UGT abstracts a proton from the substrate, allowing conjugation of substrate to glucuronic acid. Following this, protonation of the leaving group, UDP, by an acidic residue within the active site allows the release of UDP and the conjugated metabolite. Finally, proton exchange between the aforementioned basic and acidic residues within the active site reforms the active UGT molecule.

During the glucuronidation reaction a chemical inversion occurs, altering the α-UDP-glucuronic acid to a β-glucuronide conjugate. This inversion allows the phenomenon of enterohepatic recirculation to occur, which increases the half-life of some compounds metabolized via this pathway. Following Phase II metabolism, the increased polarity of the metabolite causes increased rates of excretion via the hydrophilic urine and faeces. Compounds excreted via the faeces pass into the bile and then into the intestine for excretion. However, while in the intestine chemicals are subject to further metabolism, catalysed by enzymes produced by gut bacteria: one such enzyme is β-glucuronidase. Bacterial β-glucuronidase is capable of cleaving glucuronide conjugates from chemicals, reforming the parent compound, which may then be re-absorbed into the body. As the glucuronidase produced by the bacteria in the gut is specific for the β-glucuronide then the phenomenon of enterohepatic recirculation only occurs because of the chemical inversion of α-UDP glucuronic acid to β-glucuronide conjugate during glucuronidation. *Figure 3.2* shows the enterohepatic recirculation of chloramphenicol, which results in an increased half-life in humans.

3.2.3 UDP-glucuronosyltransferase pharmacogenetics

As described in the previous section, the UGT1A gene structure is unusual, with a single locus encoding several different proteins through alternate splicing of the first exon (*Figure 3.3*). Because of this conservation of exons 2, 3, 4 and 5 it might be thought that the scope for polymorphisms within this cluster is limited. This, however, does not appear to be the case, and in fact, the reverse may well be true, as a polymorphism in any of the conserved exons affects all UGT1A enzymes within an individual, potentially exaggerating the biological significance of such a polymorphism. Mackenzie *et al.* (1997) have proposed a nomenclature to describe the UGT

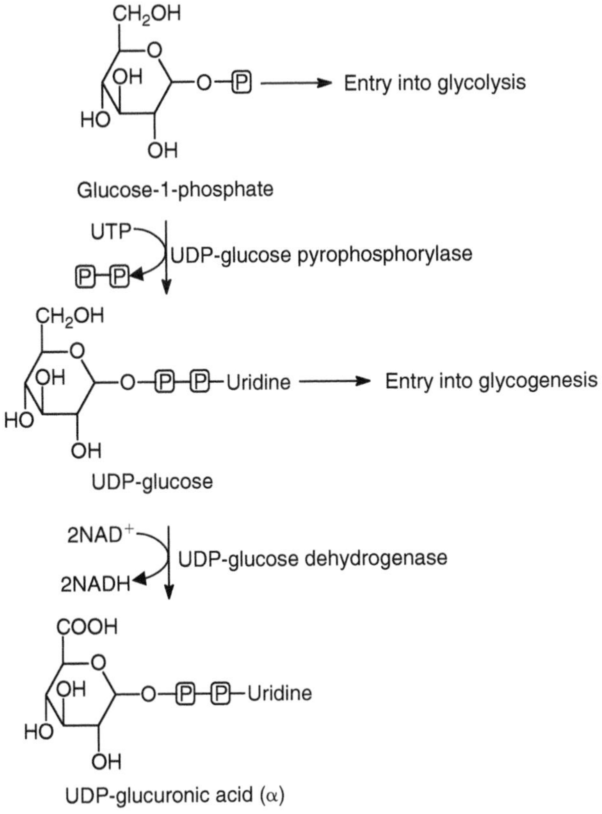

Figure 3.1

Formation of UDP-glucuronide.

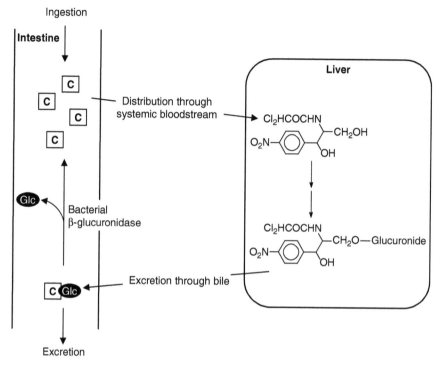

Figure 3.2

Enterohepatic recirculation of chloramphenicol.

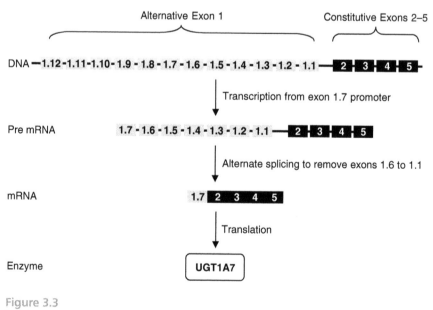

Figure 3.3

The *UGT1* locus (adapted from Tukey and Strassburg, 2000).

polymorphisms, taking into account this complicated gene structure. Some of the clinically most important polymorphisms are listed in *Table 3.3*.

Table 3.3 Clinically important polymorphisms of human UDP-glucuronosyltransferase enyzmes

Polymorphism(s)	Notes
UGT1A1*2-27	Crigler-Najjar syndrome
UGT1A1*28, *33, *34	Gilberts syndrome
UGT2B4*D^{458}	Conjugation of bile acids and sex hormones

3.2.4 UDP-glucuronosyltransferase-mediated toxicity

In general, glucuronidation reactions are thought of as detoxification reactions but a few examples do exist where glucuronidation results in a more toxic metabolite. Carboxyl-containing compounds, including both endogenous compounds (bilirubin, lithocholic acid) and xenobiotics (NSAIDs, sodium valproate) may undergo glucuronidation to form acyl glucuronides. These acyl glucuronides are then subject to 'acyl migration', involving intramolecular transesterification reactions at the C2–C4 hydroxyl groups of the glucuronic acid (*Figure 3.4*). Migration of the glucuronide from the C1 to C2 position is irreversible, while migrations between C2, C3 and C4 may continue to occur. Acyl glucuronides are chemically unstable and may become chemically reactive at pH > 7. Under these alkaline conditions these C2–C4 acyl glucuronides may cause reversible or irreversible binding to proteins, the latter probably occurring via imine formation (*Figure 3.4*; Spahn-Langguth and Benet, 1992).

3.3 Sulphate conjugation

3.3.1 Nomenclature and structure of sulphotransferases

A family of sulphotransferase enzymes (SULTs) exist to catalyse sulphate conjugation to the myriad of substrates presented to the human body. This family may be divided into two distinct classes, dependent upon the subcellular localization of the enzyme; membrane-bound enzymes or cytosolic. Membrane-bound SULTs are found predominantly within the Golgi apparatus and are responsible for the metabolism of macromolecular endogenous structures and are outside the scope of this text. The second class, and the one which will be discussed herein, is the cytosolic SULTs, which are responsible for the metabolism of many small endogenous molecules, such as neurotransmitters and xenobiotics.

At present over 45 cytosolic SULTs are known to exist across a wide variety of species, with some 11 enzymes present in man. In keeping with the common nomenclature systems used for other drug-metabolizing enzymes SULTs are divided into families according to their identity at the amino acid level, with enzymes within the same family sharing > 45% identity, rising to > 60% for enzymes within the same subfamily. Following the established nomenclature SULTs are hence named SULT, family, subfamily, individual. Cytosolic SULTs are divided into two main families: SULT1 enzymes are often termed the phenol sulphotransferases and SULT2 enzymes the steroid sulphotransferases, although this distinction by substrate type is not a hard and fast rule. Recently, a brain-specific human SULT (SULT4A1) was identified

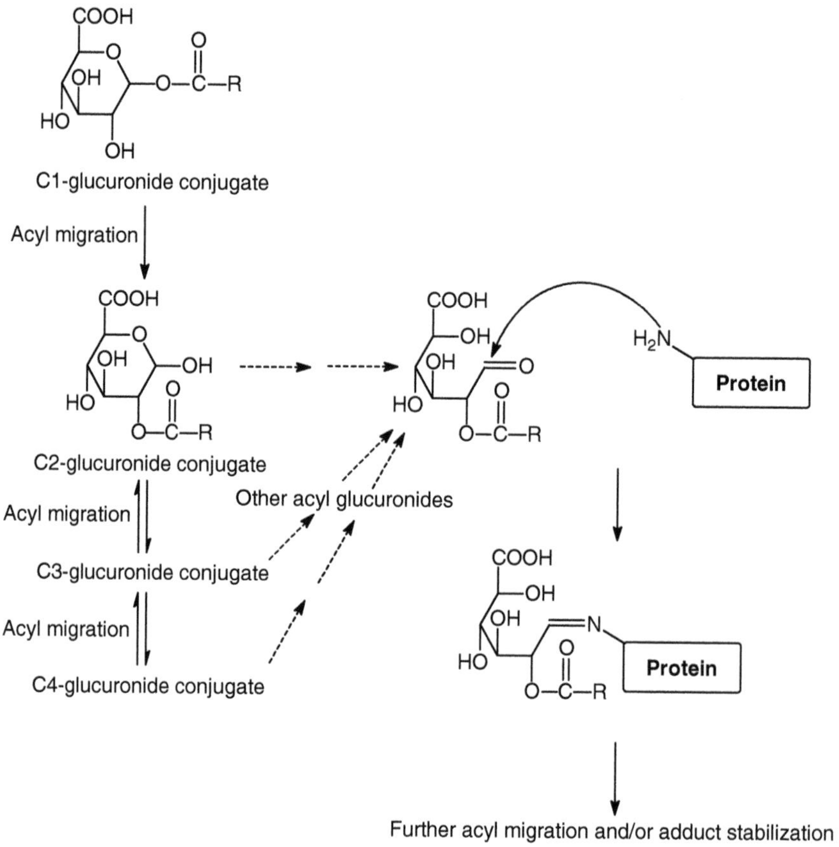

Figure 3.4

Formation of acyl glucuronides (adapted from Spahn-Langguth and Benet, 1992).

and allocated to a new family due to lack of sequence similarity with other human SULTs. The currently identified human cytosolic SULTs are shown in *Table 3.4*.

The 3-D structure of a human SULT has been determined, through the crystallization of SULT1A3 (Dajani *et al.*, 1999). The determined structure shows close similarity to the previously crystallized murine oestrogen sulphotransferase, and comparison of the two structures allowed determination of the PAPS binding region, and key amino acid residues within the active site. Due to the highly conserved nature of the SULT1 family members it is likely that the identified structural determinants are conserved throughout the family.

3.3.2 Sulphotransferase catalysed reactions

Sulphotransferases catalyse the transfer of a sulphur moiety onto the substrate, targeting —OH, —NH$_2$ and —SO$_2$NH$_2$ groups with the general stoichiometry,

$$R—OH + PAPS \rightarrow R—SO_4 + PAP$$

where R = the rest of the substrate's chemical structure and PAPS the conjugate 3'-phosphadenosine-5'-phosphosulphate.

Table 3.4   Human sulphotransferase enzymes

Name	Detoxification	Toxication	Accession
SULT1A1	Paracetamol	HMP	NM 001055.1
SULT1A2	2-Naphthol	OH—AAF, OH—APP	NM 001054.1
SULT1A3	Dopamine	Minor only	NM 003166.1
SULT1B1	1-Naphthol	HMBP, OH—CPC	NM 014465.1
SULT1C1	4-Nitrophenol	None known	NM 001056.1
SULT1C2	4-Nitrophenol	Minor only	NM 006588.1
SULT1E1	17β-oestradiol	HMP	NM 005420.1
SULT2A1	Dehydroepiandrosterone	Hycanthone	NM 003167.1
SULT2B1	Dehydroepiandrosterone	None known	NM 004605.1
SULT4A1	None known	None known	NM 01435.1

Major human SULTs are listed, along with examples of substrates metabolized in both detoxification and toxication reactions and RefSeq mRNA accession number. HMP, 1-hydroxymethylpyrene, HMBP, 6-hydroxymethylbenzo[a]pyrene; OH—CPC, 4-hydroxycyclopenta[def]chrysene, OH—AAF, N-hydroxy-2-acetylaminofluorene, OH-APP, 2-hydroxylamino-5-phenylpyradine. Adapted from Glatt et al. (2001).

In common with glucuronidation, prior to transfer of the conjugate (in this case inorganic sulphur) to the substrate, there is a need to 'activate' the conjugate by incorporating it into a chemically reactive molecule that can act as the sulphuryl donor. In the case of SULTs this donor is 3'-phosphadenosine-5'-phosphosulphate, formed by a two-step conversion of inorganic sulphate using two molecules of ATP as shown in *Figure 3.5*.

Following activation of inorganic sulphate to PAPS, transfer to substrate can occur. The reaction scheme for sulphotransferases is relatively simple with the only prerequisite being that both sulphuryl donor (PAPS) and acceptor (substrate) are present within the active site. Unlike enzymes such as CYPs or FMOs where the catalytic cycle is constrained, with co-factors and substrates interacting at set points within the catalytic cycle, binding and disassociation of PAPS and substrate into/from the active site of SULTs is essentially random and may occur in any order.

3.3.3   Sulphotransferase pharmacogenetics

Much work has been undertaken to study the presence and functional consequence of polymorphisms within the genes encoding the SULTs. Single nucleotide polymorphisms have been identified within all of the

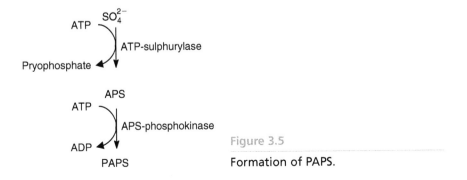

Figure 3.5

Formation of PAPS.

human SULT genes, but these are generally silent and hence of little interest in terms of response to chemical exposure. However, functional polymorphisms have been identified in both the *SULT1A1*, and *SULT1A2* genes. All of the 15 *SULT1A1* and 13 *SULT1A2* alleles so far identified result in either a silent change or decreased enzyme activity (Raftogianis *et al.*, 1999). Thus, overall, SULT polymorphisms either have no effect on drug metabolism, or decrease the sulphotransferase component of this metabolism. As sulphotransferase activity tends to only occur at low substrate concentrations, due to the limiting amount of inorganic sulphate available to the body the effect of such polymorphisms is relatively minor; the major outcome of such a decrease is usually an increased glucuronidation component to metabolism, as this is the pathway that often replaces sulphotransferase activity at higher substrate concentration anyway.

3.3.4 Sulphotransferase-mediated toxicity

Phase II metabolism involving conjugation to sulphate is responsible for the metabolic activation of a number of procarcinogens and mutagens. Because of the action of sulphate as an electron-withdrawing group in chemical structures, metabolites may be chemically unstable and undergo non-enzymatic heterolytic cleavage, resulting in an electrophilic breakdown product. Such spontaneous degradation occurs in a number of benzylic, allylic alcohols and aromatic hydroxylamines following metabolism by SULTs. The resultant breakdown products are highly electrophilic and may interact with DNA and/or proteins to form adducts.

An example of such a bioactivation reaction is provided by 2-nitropropane, an industrial solvent with widespread usage. Rats exposed to 2-nitropropane, via either oral or inhalation routes, develop liver cancer, and this has been attributed to adducts formed by highly reactive nitrenium ions, which are formed during the non-enzymatic breakdown of sulphate conjugates of 2-nitropropane (*Figure 3.6*; Sodum *et al.*, 1994). To obtain an accurate risk assessment for humans from exposure to 2-nitropropane it was therefore necessary to investigate if nitrenium ion formation could occur in humans. Kreis *et al.* (2000) expressed individual human sulphotransferases *in vitro* in V79 cells and examined their ability to cause DNA modification and induce DNA repair systems within the cells. Both SULT1A1 and SULT1C1 were shown to be capable of supporting 2-nitropropane-mediated DNA damage, and thus it may be concluded that the potential for genotoxicity by 2-nitropropane exists in humans. However, the lack of human cancers directly linked to 2-nitropropane may suggest that the risk is low, possibly because a second route of metabolism, denitrification by cytochrome P450s is the major route in humans, whereas sulphation is the major route in rodents.

3.4 Glutathione conjugation

In contrast to glucuronidation and sulphation, conjugation to glutathione utilizes a chemically inert conjugate and is targeted towards chemically reactive substrates. These substrates include alkyl or aryl halides, isothiocyanates, α, β-unsaturated carbonyls, quinines and epoxide-containing

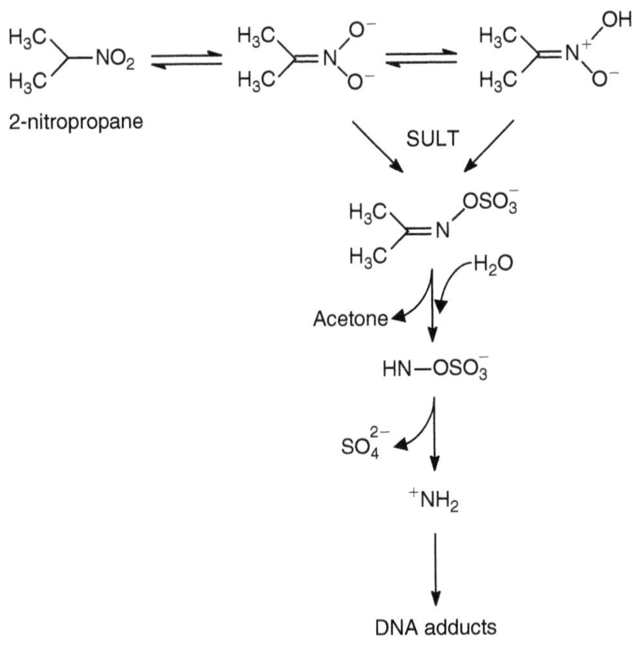

Figure 3.6

Proposed mechanism of activation of 2-nitropropane by sulphotransferases.

chemicals. Hence, glutathione conjugation is often thought of as a safety system within the cell designed to soak up any reactive metabolites produced during cellular processes, including Phase I metabolism, before they can do any cellular damage through adduct formation with cellular macromolecules.

3.4.1 Nomenclature and structure of glutathione transferases

Glutathione S-transferase enzymes (GSTs) have been identified in a large number of evolutionary diverse species, underlining the important role this class of enzymes plays in cellular defence from chemically reactive species. To date, GSTs have been identified in bacteria, insects, plants, fish, birds and mammals.

The GST enzymes have been divided into two distinct superfamilies based upon their sub-cellular localization; cytosolic or microsomal. Of these the former is the larger class, and was discovered some 30 years earlier than the recently identified microsomal GSTs, also referred to as MAPEG enzymes. The nomenclature adopted for the cytosolic GSTs is different from that used for other drug-metabolizing enzymes, using a Greek letter to represent class followed by a unique numerical identifier, although the basis for inclusion within a family, amino acid identity, is the same. In addition, *in vivo* GSTs function as either a homo- or heterodimer and the nomenclature is designed to reflect this with the enzyme GSTM1-1 being formed from a homodimer of two μ1 proteins. *Table 3.5* details the major human GST proteins from which the dimeric, functional enzymes are derived.

Table 3.5   Human glutathione *S*-transferase enzymes

Name	Expression	Accession
GSTα1	Unknown	L 13269.1[a]
GSTα2	Brain, heart, liver, prostate, kidney, lung	NM 000846.2
GSTα3	Unknown	NM 000847.2
GSTα4	Marrow, spleen, brain, heart, liver, prostate, kidney, lung	NM 001512.1
GSTμ1	Ubiquitous	NM 000561.1
GSTμ2	Brain, skeletal muscle, liver, lung	NM 000848.1
GSTμ3	Brain, heart, pancreas, prostate, kidney, lung	NM 000849.1
GSTμ4	Spleen, brain, heart, pancreas, prostate, kidney, lung	NM 000850.2
GSTμ5	Thymus, brain, heart, prostate	NM 000851.1
GSTπ1	Ubiquitous	NM 000852.2
GSTτ1	Ubiquitous	NM 000853.1
GSTτ2	Thymus, brain, heart, liver, prostate, lung	NM 000854.2
GSTζ1	Ubiquitous	NM 001513.1

Major human GSTs are listed, along with proposed distribution and RefSeq mRNA accession number. Proposed distribution is derived from expression mapping studies and data mining of EST databases. [a] GSTA1 does not have a RefSeq accession number, the NCBI accession being given instead.

3.4.2   Glutathione transferase-catalysed reactions

As stated above, GST-catalysed reactions are targeted towards chemically reactive substrates and thus do not require a chemically reactive conjugate. The major targets of this enzyme are epoxides, reactive halides and reactive oxygen radicals, but can extend to any chemically reactive group. The conjugate utilized is the tripeptide glutathione, formed from the three amino acids glutamine-cysteine-glycine. The overall reaction is electrophilic attack by the substrate on the sulphur atom of the cysteine residue of the conjugate, with the general stoichiometry shown below,

$$R—epoxide + GSH \rightarrow R—GS$$

where R = the rest of the substrate's chemical structure.

In addition to glutathione conjugation reactions, GST enzymes are capable of catalysing several other biologically important reactions, such as *cis-trans* isomerization. These reactions include the conversion of 13-*cis*-retinoic acid to all-*trans*-retinoic acid, resulting in increased affinity for the retinoid X receptor.

Another ability of GST enzymes is the catalysis of β-elimination reactions, and this property has been used in the development of a novel anti-cancer agent. An inherent problem with anticancer agents is that they are generally non-selective; they target rapidly dividing cells, including those of healthy tissues such as hair, intestine and bone marrow: the anti-cancer drug γ-glutamyl-α-amino-β(2-ethyl-*N*,*N*,*N*′,*N*′-tetrakis(2-chloroethyl) phosphorodiamidate)-sulphonyl–propionyl-(R)-(−)phenylglycine (TER-286) however uses GST-catalysed β-elimination to increase its specificity for cancerous tissue. TER-286 is delivered as a latent prodrug, which is only converted to the active nitrogen mustard derivative following β-elimination catalysed by GST (Rosario *et al.*, 2000). This derivative spontaneously rearranges to form an aziridium ring which alkylates DNA, leading to eventual cell destruction. Specificity for cancer cells is gained through the fact that GSTP enzymes are over-expressed in many human tumours (Schippet, 1997). TER-286 is currently in Phase II clinical trials.

3.4.3 Glutathione transferase pharmacogenetics

In humans there are a number of clinically relevant polymorphisms of the *GST* genes, and these are present at relatively high frequencies within the population. Due to this high frequency of clinically relevant polymorphisms it is likely that genetic variation in *GST* genes plays a significant role in the individual response to chemicals, be they therapeutic agents or toxic chemicals.

Both GSTM1 and GSTT1 have null alleles (GSTM1*0 and GSTT1*0 respectively) caused by the deletion of the entire gene. Such null alleles obviously lead to a decrease in the expression of GSTM1 and this would markedly affect the metabolism of any compound that is a substrate for GST enzymes formed from either of these monomers. As GSTs are responsible for the detoxification of a large number of chemically reactive compounds it is perhaps not surprising therefore that these null phenotypes have been associated with increased risks of cancer (Gawronska-Szklarz *et al.*, 1999; Zhang *et al.*, 1999). Several studies have linked the presence of these genotypes with an increased risk of colon cancer, presumably through the lack of detoxification of ingested procarcinogens and carcinogens. However, this area is still controversial with other researchers failing to find an association (Ye and Parry, 2002). Further research is required to examine if these null alleles really do cause a significantly increased risk of colon cancer, or whether other factors (such as the level of other GSTs in the body) play a determining role.

3.4.4 Glutathione transferase-mediated toxicity

While, as described above, GST activity is seen as a mechanism of detoxifying chemically reactive species, a number of glutathione conjugates are more reactive than the original substrate, resulting in toxication rather than detoxification reactions. These chemicals can be divided into three broad classes, haloalkanes, haloalkenes and hydroquinones (*Table 3.6*), each of which will be discussed below.

Many industrial chemicals contain halogenated alkane moieties in their structures and these are open to metabolism by GSTs. One such chemical is tris (2,3-dibromopropyl)phosphate, a chemical widely used as a flame-retardant treatment for materials. However, rats exposed to this compound are subject to kidney damage following acute exposure and renal carcinogenicity following chronic exposure. What then is the risk to humans associated with occupational exposure? To answer this question we must examine the mode of toxicity. The parent compound, tris (2,3-dibromopropyl) phosphate, is non-toxic and requires metabolic activation, initially via CYP-mediated Phase I metabolism, followed by conjugation to glutathione. Evidence that the glutathione conjugation reaction plays the key role in the observed toxicity was provided by Simula *et al.* (1993) using recombinant human GSTs. *Salmonella typhimurium* were engineered to express human

Table 3.6 Glutathione *S*-transferase-mediated toxication

Chemical class	Example
Halogenated alkanes	Dichloromethane
Halogenated alkenes	Tetrachloroethene
Isothiocyanates	Benzyl isothiocyanate

GSTs and the mutagenic rate examined following exposure to tris (2,3-dibromopropyl) phosphate. For toxicity to be exerted cultures had to be incubated with rat liver microsomes, underlining the requirement for an initial Phase I activation reaction. However, following Phase I-mediated metabolic activation, bacterium expressing either recombinant GSTP1-1 or GSTA1-1 showed a higher rate of mutagenicity than bacterium without human GSTs (25-fold increase in GSTP1-1-expressing cells). It was proposed that the glutathione conjugate undergoes spontaneous ring formation to form a highly reactive episulphonium ion, and this ion reacts with DNA/protein to form adducts within the kidney (*Figure 3.7*).

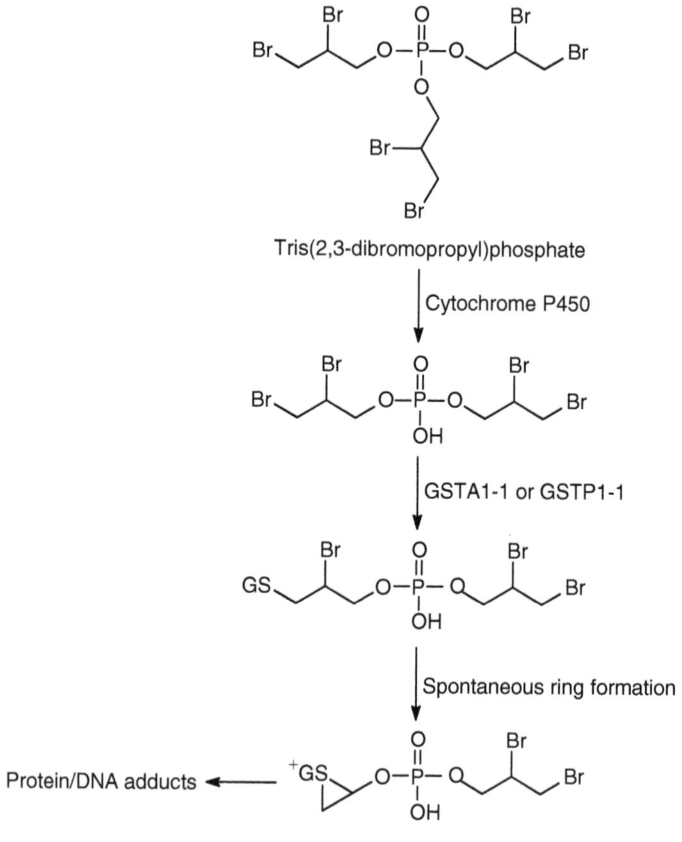

Figure 3.7

Toxicity of tris(2,3-dibromopropyl)phosphate.

As there is undoubted human exposure to haloalkanes, and it appears that human GSTs may catalyse the toxication reactions, is the associated human risk high? While such a question can really only be answered on a case-by-case basis, examination of the toxicity of dichloromethane (DCM) provides some general rules.

DCM was a widely used industrial solvent but the discovery that it was a hepatic and pulmonary carcinogen by inhalation in mice led to its classification as a possible human carcinogen and resulted in the restriction of its use. In mice two major metabolic routes exist for DCM. The first is an oxidation by CYP2E1, followed by glutathione conjugation, whereas the second involves direct conjugation of DCM to glutathione. The CYP-mediated metabolic pathway is low capacity and easily saturated, meaning that high dose or prolonged exposure leads to direct conjugation of DCM to glutathione. As with other haloalkanes this conjugate may then form a highly reactive episulphonium ion which reacts with DNA and/or protein to form adducts. In mice, the direct conjugation of DCM is catalysed by GST T1-1, which is highly expressed in the liver. Sherratt *et al.* (2002) compared the properties of murine GST T1-1 with those of its human orthologue. While both enzymes were capable of directly reacting with DCM, the K_{cat}/K_m for human GST1-1 was only 20% that of the murine GST T1-1. In addition, levels of GST T1-1 were approximately five times higher in mouse liver than in human liver. The combination of low K_{cat}/K_m and low expression levels means that the risk to humans from DCM is considerably lower than in rodents. Hence, the risk to humans from haloalkanes may be limited due to the lower efficiency of human enzymes at catalysing the pertinent reaction and reduced expression of human GSTs at the site of toxicity seen in rodents.

Haloalkenes are also subject to bioactivation by glutathione, although by a completely different route. In this case, the glutathione conjugate itself does not form the reactive species; it is the products of further metabolism that are important. The glutathione conjugate may be broken down by sequential removal of the glutamine and glycine residues of this tripeptide, to leave just the cysteine residue. Acetylation of this residue then results in the formation of the mercapturic acid derivative of the cysteine conjugate, which is then excreted in the urine (*Figure 3.8*). Instead of acetylation, however, the cysteine conjugate may be metabolized by the enzyme cysteine conjugate β-lyase (CS-lyase), removing the cysteine residue to leave a highly reactive metabolite, which may form DNA/protein adduct (*Figure 3.8*). CS-lyase is present in several tissues within the body, but is most prevalent in the kidney and brain. Rats exposed to halogenated alkenes such as the industrial solvent perchloroethene are subject to massive kidney damage, due to the formation of this reactive metabolite.

If rats are subject to such gross toxicity and halogenated alkenes are in wide use as solvents, anaesthetics, etc., what then is the risk to humans? Human occupational exposure does not appear to result in such devastating renal toxicity, although some kidney malformations have been reported. Instead, compound exposure appears to be mainly targeted to the brain, resulting in the potential for neural lesions. However, the concentration of compound and enzyme within the brain appears to be such that adverse effects are minor at most, with little clinical evidence existing to suggest a real risk to human health. Despite this some halogenated alkenes have been withdrawn from use; for example perchloroethene is no longer used as a solvent in dry cleaning factories.

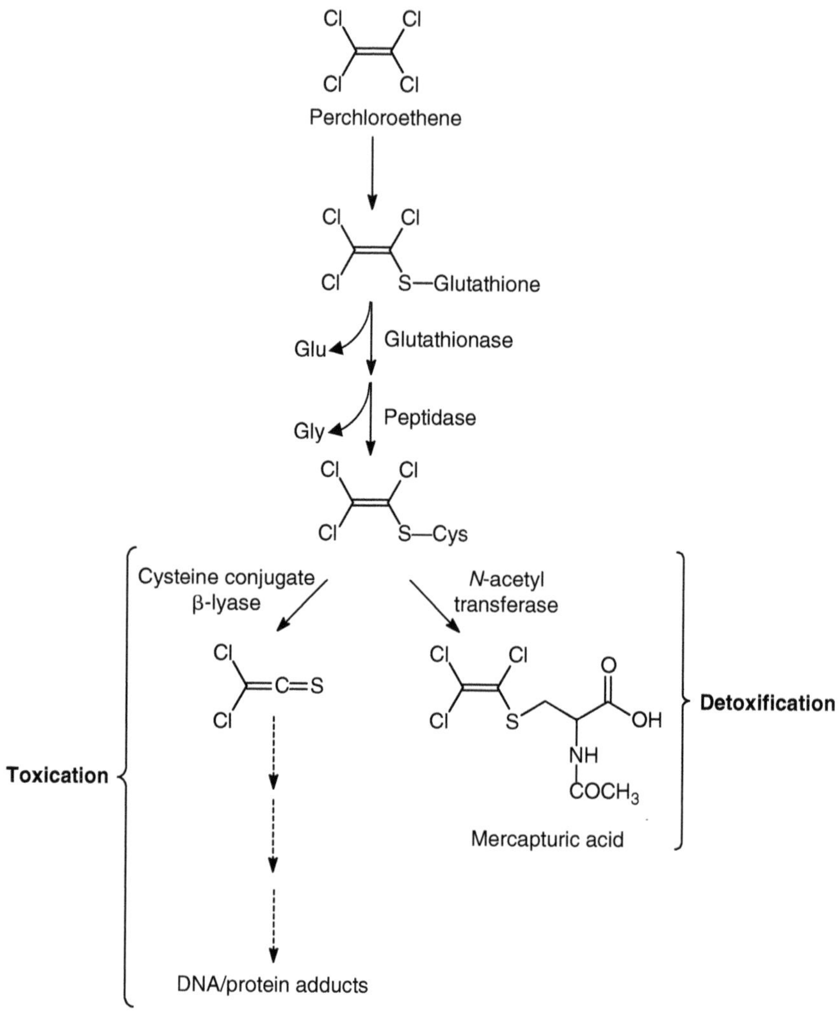

Figure 3.8

Toxicity of perchloroethylene.

3.5 Epoxide hydrolase-mediated toxicity

While the three enzyme systems discussed above represent a significant proportion of Phase II reactions, as shown in *Table 3.1* other reactions do occur. One enzyme that has not been discussed yet, but is worthy of note due to its role in both detoxification and toxication reactions is epoxide hydrolase.

In general, metabolic processes move to increase polarity and decrease toxicity. However, these two aims are not always mutually compatible, and often the introduction of functional groups during Phase I reactions results in metabolic activation of the compounds, as can be seen from the examples presented in Chapter 2. The body has a number of enzyme systems designed to cope with the production of such potentially dangerous metabolites, or indeed compounds that are chemically active without the need for metabolic activation. As we have already seen, the glutathione S-transferase enzymes are responsible for a significant proportion of these detoxification reactions. However, epoxide hydrolase also plays an important role in detoxification; the conversion of highly reactive epoxide moieties to less reactive diols (*Figure 3.9*). This reaction occurs for both endogenous epoxides, such as androstene oxide, and xenobiotics, although the enzyme appears to have a much higher affinity for the former group of chemicals.

Epoxide hydrolase can however play a role in toxication reactions. As can be seen from *Figure 3.9*, epoxide hydrolase is capable of converting the 7,8-epoxide of Benzo(*a*)pyrene, an environmental pollutant, to the diol, which is in itself non-toxic. However, this metabolite may now be subject to further metabolism by CYPs, forming the potent carcinogen benzo(*a*)pyrene-7,8-diol-9,10-epoxide.

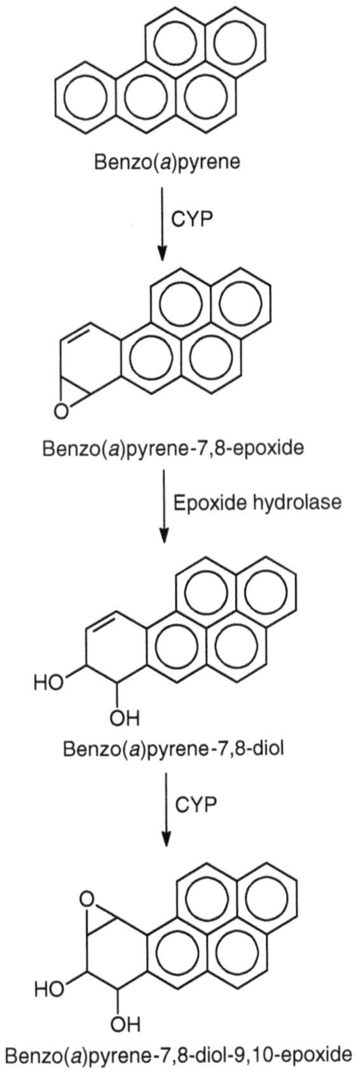

Benzo(*a*)pyrene

CYP

Benzo(*a*)pyrene-7,8-epoxide

Epoxide hydrolase

Benzo(*a*)pyrene-7,8-diol

CYP

Benzo(*a*)pyrene-7,8-diol-9,10-epoxide

Figure 3.9

Epoxide hydrolase-mediated activation of benzo[a]pyrene (adapted from Gibson and Skett, 2001.

3.6 Summary

Phase II metabolism completes the process begun with Phase I metabolism; increasing the polarity of chemicals to a point where excretion in hydrophilic medium, such as the urine and faeces, is highly favoured. Thus, the combination of Phase I and II metabolism plays a vital role in maintaining the body's homeostasis, through the removal of both endogenous and exogenous chemicals, allowing careful regulation of their levels.

In addition to completing the processing of chemicals and thus facilitating their excretion, Phase II metabolism also plays a vital role in protecting the body from toxic chemicals. Phase II metabolism rapidly metabolizes any reactive intermediates produced during Phase I metabolism and therefore removes them from the body before they can react with intracellular macromolecules and potentially cause harm to the cell. Because of this we can see that the levels of Phase I and II enzymes must be carefully balanced, as a deficit of Phase II enzymes compared to Phase I may result in the production of such reactive intermediates more rapidly than the Phase II enzyme systems can remove them. This therefore leads to a close coordination of the expression of Phase I and II genes, with similar chemicals causing increases in both sets. In the next chapter we will examine how the cell produces such coordinated responses to chemical insult.

Secondly, the Phase II glutathione S-transferases play an important role in scavenging reactive molecules within the cell, and hence act as 'biological hoovers' attempting to protect the cell from any chemically reactive species. In the next chapter we will examine how these protective mechanisms are called upon to combat another potential threat to the cell, reactive oxygen and nitrogen species.

Finally, the next chapter will deal with the most important decision a cell may have to make, whether the damage sustained is recoverable from, or whether the cell should die, and if so by what mechanism.

References

Bock, K.W. (2001) UDP-Glucuronosyltransferases. In: Ioannides C. (ed.) *Enzyme Systems that Metabolise Drugs and Other Xenobiotics*. John Wiley & Sons, Chichester, pp. 281–318.

Dajani, R., Cleasby, A., Neu, M., et al. (1999) X-ray crystal structure of human dopamine sulfotransferase, SULT1A3. *J. Biol. Chem.* **274**(53): 37862–37868.

Gawronska-Szklarz, B., Lubinski, J., Kladny, J., et al. (1999) Polymorphism of GSTM1 gene in patients with colorectal cancer and colonic polyps. *Exp. Toxicol. Pharmacol.* **51**(4–5): 321–325.

Glatt, H., Boeing, H., Engelke, C.E., et al. (2001) Human cytosolic sulphotransferases: genetics, characteristics, toxicological aspects. *Mutation Res.* **482**(1–2): 27–40.

Kries, P., Brandner, S., Coughtrie, M.W.H., et al. (2000) Human phenol sulphotransferases hP-PST and hM-PST activate propane 2-nitronate to a genotoxicant. *Carcinogenesis* **21**(2): 295–299.

Mackenzie, P.I., Owens, I.S., Burchell, B., et al. (1997) The UDP-glucuronosyltransferase gene superfamily: Recommended nomenclature update based on evolutionary divergence. *Pharmacogenetics* **7**: 255–269.

Meech, R. and Mackenzie, P.I. (1997) UDP-Glucuronosyltransferase, the role of the amino terminus in dimerisation. *J. Biol. Chem.* **272**: 26913–26917.

Raftogianis, R.B., Wood, T.C. and Weinshilboum, R.M. (1999) Human phenol sulphotransferase *SULT1A2* and *SULT1A1*. *Biochem. Pharmacol.* **58**: 605–616.

Ritter, J.K., Chen, F., Sheen, Y.Y., *et al.* (1992) A novel complex locus *UGT1* encodes human bilirubin, phenol, and other UDP-glucuronosyltransferase isoenzymes with identical carboxyl termini. *J. Biol. Chem.* **267**: 3257–3261.

Rosario, L.A., O'Brien, M.L., Henderson, C.J., *et al.* (2000) Cellular response to a glutathione S-transferase P1-1 activated prodrug. *Mol. Pharmacol.* **58**: 164–174.

Schipper, D.L. (1997) Glutathione *S*-transferases and cancer. *Int. J. Oncol.* **10**: 1261–1264.

Sherratt, P.J., Williams, S., Foster, J., *et al.* (2002) Direct comparison of the nature of mouse and human GST T1-1 and the implications on dichloromethane carcinogenicity. *Toxicol. Appl. Pharmacol.* **179**(2): 89–97.

Simula, T.P., Gluncey, M.J., Soderlund, E.J., *et al.* (1993) Increased mutagenicity of 1,2-dibromo-3-chloropropane and tris(2,3-dibromopropyl)phosphate in *Salmonella* TA100 expressing human glutathione *S*-tranferases. *Carcinogenesis* **14**: 2303–2307.

Sodum, R.S., Sohn, E.S., Nie, G., *et al.* (1994) Activation of the liver carcinogen 2-nitropropane by aryl sulphotransferase. *Chem. Res. Toxicol.* **7**: 1947–1949.

Spahn-Langguth, H. and Benet, L.Z. (1992) Acyl glucuronides revisited: is the glucuronidation process a toxification as well as a detoxification mechanism? *Drug Metabol. Rev.* **24**(1): 5–47.

Tukey, R.H. and Strassburg, C.P. (2000) Human UDP-glucuronosyltransferases: metabolism, expression, and disease. *Ann. Rev. Pharmacol. Toxicol.* **40**: 581–616.

Ye, Z. and Parry, J.M. (2002) Genetic polymorphisms in the cytochrome P450 1A1, glutathione *S*-transferase M1 and T1, and susceptibility to colon cancer. *Teratogenesis Carcinogenesis and Mutagenesis* **22**(5): 385–392.

Zhang, H., Ahmadi, A., Arbman, G., *et al.* (1999) Glutathione *S*-transferase T1 and M1 genotypes in normal mucosa, transitional mucosa and colorectal adenocarcinoma. *Int. J. Cancer* **84**(2): 135–138.

Co-ordinated responses to toxicity

4

4.1 Introduction

In the previous two chapters we have described the body's main systems for regulating levels of both xenobiotics and endogenous compounds; these being the Phase I and II metabolic enzymes. These systems are very efficient in dealing with the low levels of exposure to chemicals that the body is usually presented with. However, what happens when exposure is at higher levels? The levels of Phase I and II enzymes can be induced to respond to high levels of compound, providing a responsive system designed to adjust to the fluctuating levels of chemicals to which the body is exposed. However, this response may not always be optimal. Firstly, there are limits to the capacity for induction of these enzymes and hence a point will come where more chemical does not result in more enzymes. Secondly, induction of all the enzymes involved in a chemical's safe and efficient metabolism may not occur to the same degree, resulting in an uneven ability of the cell to deal with a chemical and/or its metabolites. Under such circumstances, the level of a chemical will outweigh the ability of the body to safely eliminate it – at this point a toxic response may occur if other systems are not activated. In this chapter we will look at the series of events that occur within a cell to cope with such an overload, and either remove the toxic threat or dispose of damaged cells.

4.2 Immediate responses to toxic insult

A surprising aspect of metabolism is that the very systems designed to alter chemicals to enable their safe and efficient removal from the body may also produce a number of chemical species which are toxic to the cell (i.e. bioactivation). These species may be divided into two broad categories, reactive intermediates formed from a chemical during its metabolism (i.e. large chemical structures with chemically reactive moieties within them) and small, chemically reactive species released during metabolism (i.e. small mono- or di-atomic species).

4.2.1 Generation of small, chemically reactive species

Many chemicals within the body are capable of exchanging electrons, thus altering their oxidation status. Oxygen atoms are particularly susceptible to reduction via the gain of between one and four electrons, resulting in the creation of reactive oxygen species (ROS), which are potent oxidants (*Figure 4.1*). The release of ROS will result in a change in the reduction: oxidation (redox) potential within the cell, and this results in the phenomenon known as oxidative stress. During oxidative stress the ROS molecules generated can carry out nucleophilic attack on any electron-deficient chemical group unless removed rapidly. Potential targets include most large cellular macromolecules, such as protein, lipid and DNA, and result in the formation of adducts, covalent binding of the ROS to the macromolecule, which disrupts cellular functions.

As many metabolic pathways use oxygen, in the form of diatomic oxygen (O_2), it may be thought that we would be constantly at risk of forming potent oxidants through the sequential gain of electrons by oxygen. However, in the majority of these processes, such as during respiration, the sequential reduction of O_2 is closely controlled and ROS are not released into the cell. Instead all four, single electron reductions, are carried out sequentially, forming H_2O, and this can then be safely excreted (*Figure 4.1*).

As discussed in Chapter 2, diatomic oxygen (O_2) is a key component of many Phase I reactions, in particular the reactions catalysed by the cytochrome P450s. During the CYP catalytic cycle diatomic oxygen becomes bound to the ferrous haem moiety of CYP, in a similar fashion to that seen in the binding of oxygen to the haem group of haemoglobin. Electron transfer between the haem group and diatomic oxygen results in the formation of superoxide and then peroxide species. The highly reactive peroxide then reacts with the substrate, resulting in the production of oxidized metabolite and water (*Figure 4.2*). This reaction therefore results in the stabilization of the reactive peroxide intermediates through their incorporation into both the substrate and water. However, in the absence of a suitable substrate, it is possible that ROS may be released before O_2 has been fully reduced to water.

If ROS formation is only a problem when no recipient for the ROS exists (i.e. during uncoupled reactions) then one simple way of preventing oxidative stress is to keep iron and oxygen separate until all other constituents of the reaction cycle are present, and this can be achieved in two ways. Firstly, molecular iron is stored within the cell as ferritin and transported as transferrin and its sequestration in these molecules effectively makes it metabolically unavailable, preventing interaction with intracellular oxygen. Secondly, even once iron has been incorporated into complete haemoproteins, it is still protected from undesirable interactions with diatomic oxygen by the 3-dimensional structure of the haemoprotein. Haemoprotein active sites are such that the prosthetic haem is hidden within the heart of the protein, and hence the chances of 'accidental' uncoupled reactions between iron and diatomic oxygen are greatly reduced. Binding of substrate causes a conformational change that allows easier access for O_2 to the prosthetic haem.

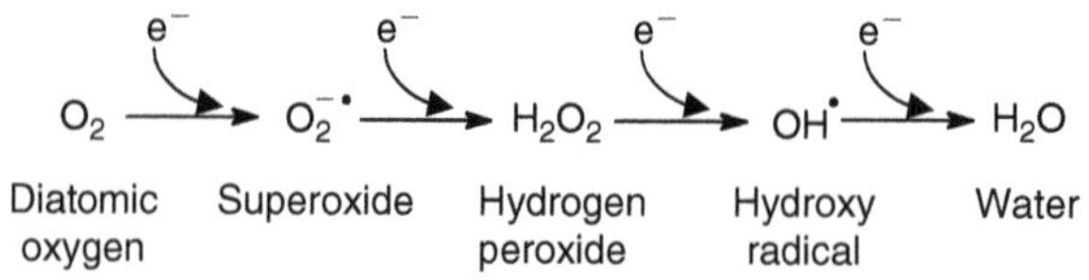

$$O_2 \xrightarrow{e^-} O_2^{-\bullet} \xrightarrow{e^-} H_2O_2 \xrightarrow{e^-} OH^\bullet \xrightarrow{e^-} H_2O$$

| Diatomic oxygen | Superoxide | Hydrogen peroxide | Hydroxy radical | Water |

Figure 4.1

Generation of reactive oxygen species.

$$Fe(II) + O_2 \rightleftharpoons Fe(III) + O_2^{-\bullet} \quad \text{(Superoxide)}$$

$$Fe(III) + O_2^{-\bullet} \rightleftharpoons Fe(III) + O_2^{2-} \quad \text{(Peroxide)}$$

Figure 4.2

Reactive oxygen species generated during the cytochrome P450 catalytic cycle.

Despite these precautions, uncoupled reactions may occur, and result in the release of ROS. For example, if a substrate is difficult to oxidize, then uncoupling of the substrate may occur before stabilization of the ROS has happened. In these situations, reactive superoxide and peroxide will be released into the cell. This situation may be exacerbated since the uncoupled substrate will cause induction of CYP protein levels; this in turn will result in an increased production of ROS. Such situations are particularly evident with the CYP2E family of CYPs, as these enzymes naturally exist in a high spin state, increasing the possibility of uncoupled reactions occurring. Sakurai and Cederbaum (1998) demonstrated that liver cells engineered to over-express CYP2E1 were more susceptible to the toxicity of ferric-nitriloacetate, caused by excessive generation of ROS. This toxicity led to lipid peroxidation and cell death via apoptosis, and was further exacerbated by the depletion of glutathione, one of the body's key defence mechanisms against ROS.

As stated in the introduction, production of such highly reactive species at high concentration may be deleterious to the cell. To prevent the formation of adducts the cell has a number of enzymes whose function is to remove ROS before they can cause adducts. *Table 4.1* details the main enzymes involved in these systems and the reactions that they catalyse. As can be seen from *Table 4.1* the role of these enzymes is to complete the reduction of ROS, forming water, which can then be safely from the cell.

Section 4.3.1 will discuss how the cell uses the production of ROS as an internal messenger system to alter its gene expression and thus up-regulate the expression of these genes, plus others that are required to prevent and/or recover from ROS-mediated adduct formation.

4.2.2 Generation of larger chemical structures with reactive groups

Cellular responses to chemically reactive species have been extensively discussed in Chapters 2 and 3, and so will be covered only minimally herein. The basic mechanism for safe removal of larger reactive chemicals is the same as for the removal of ROS – further chemical reactions to produce a non-reactive compound. Such metabolism may be catalysed by several of the Phase II enzymes, although the major group is the glutathione transferases.

Table 4.1 Mechanisms for detoxification of reactive oxygen species

Enzyme system	Chemical reaction
Superoxide dismutase	$2O_2 + 2H^+ \rightarrow H_2O_2 + O_2$
Catalase	$2H_2O_2 \rightarrow 2H_2O + O_2$
Glutathione peroxidase	$2GSH + H_2O_2 + 2H^+ \rightarrow GSSG + 2H_2O$

GSH, glutathione; GSSG, reduced glutathione. Adapted from Ryan and Aust (1992).

As discussed in Chapter 3, because of the low chemical reactivity of the glutathione conjugate, it tends to target towards interactions with highly reactive chemicals, and in doing so detoxifies them. Because of this, glutathione transferases are often thought of as 'biological hoovers', whose role is to clean up chemicals within the cell that may be of harm to the cell.

Glutathione transferase enzymes are inducible by a range of chemicals, as are all Phase I and II enzymes. This therefore provides a safety net system that allows up-regulation of enzyme levels in the presence of increased levels of reactive chemical, ensuring fast and efficient removal of the potentially toxic chemical. Hence, the system is able to adapt to changing levels of substrate. Toxicity may occur, however, under two circumstances. Firstly, the degree of induction is not limitless, and there will exist a point beyond which the amount of reactive chemical present cannot be removed efficiently by glutathione transferases. In the second situation, uneven induction of enzymes is caused by the chemical. This results, for example, in high induction of the Phase I enzyme producing the reactive chemical but only poor induction of the Phase II glutathione transferase removing the reactive chemical. Such imbalances will lead to an accumulation of reactive chemical, often leading to toxicity. Examples of both situations are given in Chapter 3.

4.3 Co-ordination of the response to reactive chemicals

Throughout this text exposure to chemicals is said to elicit specific changes in a cellular environment, many of which are covered within this chapter. This is usually caused by altering gene expression within the cell so that genes whose products are designed to deal with the toxic insult are switched on whereas the expression of other genes is switched off. Therefore, a pertinent question is '*How* does this overall change in cellular environment occur?' Two distinct mechanisms are important in controlling the cellular response to toxic and oxidative stress: direct and indirect activation of gene expression.

Chemicals/ROS themselves may activate intracellular receptors that directly regulate transcription of target genes, and this represents a *direct* activation of gene expression. These molecular mechanisms of transcription factor interaction with DNA, and how this affects gene expression will be extensively reviewed in Chapter 7.

Alternatively, the chemicals/ROS may interact with other molecules within the cell, and it is these molecules that propagate the signal and produce the concerted cellular response to the chemical exposure; this is hence an *indirect* activation of gene transcription by the toxic stimuli.

Indirect activation of gene transcription is propagated through signal transduction pathways, and these are an important part of the overall co-ordination of cellular response. Such pathways are capable of responding to a large range of stimuli, including toxic and oxidative stress, and elicit a range of biological responses including cell differentiation, cell movement, cell division and cell death. Perhaps the best studied of these transduction pathways are the mitogen-activated protein kinase (MAPK) cascades (Schaeffer and Weber, 1999). The importance of MAPK cascades for co-ordination of cellular responses to stress was recently demonstrated by the identification of a MAPK cascade in the nematode, *C. elegans*, showing the high degree of evolutionary conservation for this signalling pathway (Kim *et al.*, 2002).

MAPK signalling systems exist as a three-tiered cascade, with each cascade being involved in the co-ordination of response to specific stimuli. Within each cascade the biological response is mediated by a MAPK. However, before this enzyme can carry out its functions it must first be activated, via phosphorylation, by a MAPK kinase (MAPKK). This, in turn, is activated through phosphorylation by a MAPKK kinase (MAPKKK). Activation of the MAPKKK is caused by the presence of the stimulus. *Figure 4.3* shows the MAPK cascades identified in humans, the stimuli they respond to and the resultant cellular changes they elicit.

How then does an activated MAPK bring about the cellular changes seen in *Figure 4.2*? As its name suggests, MAPK acts via phosphorylating target proteins, and this can either activate or repress their biological activity. The target proteins for MAPK are often transcription factors, or co-activators/co-repressors and their phosphorylation is an important step in the formation of the active transcription factor complex that binds to the DNA and causes gene transcription to occur.

4.3.1 Small reactive chemical species as signalling molecules

The consequences of ROS production are twofold. Firstly, the reactive species may react with cellular macromolecules in their vicinity, forming protein or DNA adducts. Such a reaction is undesirable as it results in gross cellular damage, and at high enough levels this inevitably leads to cell destruction. However, at substantially lower levels, ROS act as intracellular signalling molecules. ROS fulfil the major criteria of signalling molecules in that there are several, generally controllable, mechanisms for their production and subsequent biological removal from the cell, and that these small molecules are capable of diffusing short distances within the cell.

Once produced, ROS are capable of affecting a plethora of biological processes, from cell proliferation to apoptosis. In addition, these effects can be in both directions, with ROS capable of both stimulating cell division and inhibiting it (Morel and Barouki, 1999). To study the extent that ROS can

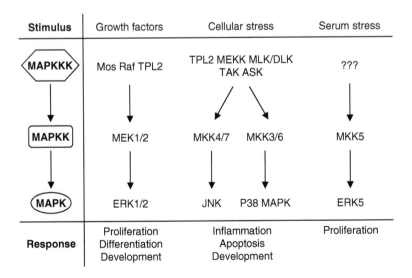

Figure 4.3

Mammalian MAP kinase cascades (adapted from Schaeffer and Weber, 1999).

affect levels of gene expression DNA microarray technology can be used to examine the changes in the transcriptome during oxidative stress (see Chapter 8 for a description of this technology). Chuang *et al.* (2002) studied the changes in gene expression in mammary cells following exposure to three oxidants, H_2O_2, menedione and t-butyl hydroperoxide, using a microarray of 17 000 genes. They identified 421 genes that were altered by all three treatments, and showed the overall pattern of gene expression was very similar regardless of the ROS source. Using such technologies not only identifies those genes that are central to ROS-mediated gene expression regardless of oxidant source, but also identifies novel genes previously not associated with ROS-mediated responses. This work and that of others has allowed a picture of the cellular changes caused by ROS to be hypothesized, showing how ROS exposure results in so many different cellular fates (*Figure 4.4*).

Once we have established that ROS are capable of causing large-scale alterations in the gene expression of cells, we need to study how this is achieved. The first possible mechanism would involve direct activation of transcription factors by ROS. Direct interaction of ROS with transcription factors has not yet been demonstrated, although circumstantial evidence for such activation has been shown. Certain cysteine residues in proteins have been

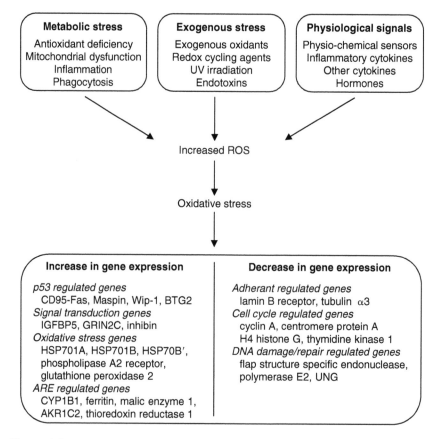

Figure 4.4

Gene expression changes caused by reactive oxygen species (data from Morel and Barouki, 1999 and Chuang *et al.*, 2002).

postulated to react with ROS, producing conformational changes in the target protein, and altering their function. Itoh *et al.* (1999) identified a protein, Keap1, which contained several such cysteine residues, and which acts as a negative regulator of the transcription factor Nrf2. They hypothesized that ROS activation of these residues resulted in the disassociation of Keap1 from Nrf2, and thus activates this transcription factor to cause changes in gene expression. Nrf2 interacts with both the antioxidant responsive element (ARE) and the electrophile responsive element (EpRE), which are found in the promoters of the genes for several Phase II enzymes and anti-oxidative stress proteins, such as superoxide dismutase and catalase. Hence, ROS may directly interact with a transcriptional co-repressor protein, causing disassociation from a transcription factor that then activates transcription of genes involved in the oxidative stress response.

In comparison to the putative direct activation of gene transcription caused by ROS interaction with Keap1, indirect activation of gene transcription by ROS is an established phenomenon. As can be seen from *Figure 4.4*, many different stimuli result in the generation of ROS, and therefore oxidative stress. Through the co-ordination of these many different stimuli into a single cellular phenomenon, ROS production and oxidative stress, allows a general stress response to be achieved to a vast array of divergent stress stimuli (Morel and Barouki, 1999). How does this state of oxidative stress indirectly affect gene expression, and how can these cause different effects in different cell types? Guyton *et al.* (1996) showed that a range of mammalian cell lines were 'activated' by ROS, and this exposure resulted in an up-regulation of several MAPK proteins. As we have already seen, MAPK cascades play a key role in regulating gene expression, and hence many cellular processes. Activation of these cascades therefore provides a mechanism by which indirect activation of gene expression by ROS could occur. Guyton *et al.* also showed that some cell lines expressing constitutively active MAPK proteins were naturally more resistant to H_2O_2 toxicity that those with normal, or defective MAPK cascades. This suggested another facet of the activation of MAPK cascades by low concentrations of ROS; through the resultant changes in gene expression the cell becomes better able to cope with the potentially dangerous adducts formed by interaction of ROS with cellular macromolecules.

In addition to the cell-signalling role of reactive oxygen species described above, the body also utilizes reactive nitrogen species for cellular signalling, and these species may impact upon ROS-mediated toxicity. Nitric oxide synthase (NOS) enzymes are capable of catalysing the five electron oxidation of one of the equivalent guanido nitrogens of L-arginine to yield L-citrulline and nitric oxide (*Figure 4.5*). In humans there are three NOS enzymes, termed eNOS, nNOS and iNOS (*Table 4.2*). The first two enzymes are named after the tissue they were first identified in (epithelium and neuronal tissue respectively), although they have since been shown to have a more wide-ranging expression. Both eNOS and nNOS are sometimes referred to as cNOS, as their expression is constitutive. By comparison, the final enzyme, iNOS, is normally present at negligible levels but is inducible by many immuno-stimulants such as inflammatory cytokines (i.e. TNFα, IFNγ and IL1/2). iNOS therefore plays an important role in host-defence against pathogens and/or chemicals that cause a pro-inflammatory response. As was seen with ROS, NO may act as a signalling molecule and causes similar, wide-ranging, changes in gene expression as seen with ROS. These effects are particularly evident in situations where iNOS levels are increased (Beck *et al.*, 1999).

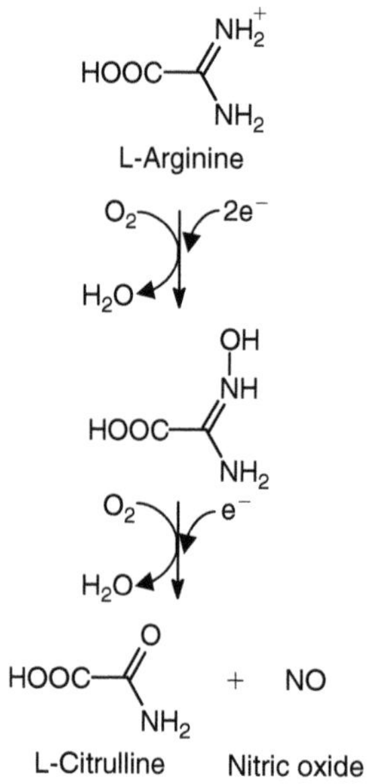

Figure 4.5 Production of nitric oxide.

The use of two different sets of reactive species as secondary messengers creates a potential problem, as the body must balance the signals generated by these two, inter-related pathways. NO may react with ROS to form multiple reactive nitrogen species, in particular the highly reactive peroxynitrite (ONOO⁻; *Figure 4.6*), and these have been associated with high levels of necrosis (Liebermann *et al.*, 1995). Hence, to achieve a co-ordinated response to stimuli the cell requires a balanced production of NO and ROS. Sumbayev *et al.* (2002) studied the inter-relationship between these two signalling pathways, and compared their regulation in smooth muscle and liver cells. They showed that in smooth muscle cells NO production was antagonistic to ROS signalling, and prevented the apoptosis usually associated with ROS expression in muscle cells. In comparison, in liver cells production of both ROS and NO resulted in high levels of apoptosis. Further investigation showed that the difference between the two cell lines was their respective levels of glutathione, one of the primary defence mechanisms against chemically reactive species (*Table 4.1*). In liver cells, the lower levels of GSH were rapidly depleted as they reacted with peroxynitrite formed through the interaction of ROS and NO, resulting in cellular damage and, ultimately, increased apoptosis. In comparison, in muscle cells GSH levels were high enough to remove the peroxynitrite caused by the interaction of NO and ROS. Hence, no ROS, NO or peroxynitrite was available to cause cellular damage and therefore apoptosis.

Table 4.2 Human nitric oxide synthase enzymes (NOS)

Enzyme	Expression	RefSeq accession
nNOS (NOS1)	Constitutive	NM 000620.1
iNOS (NOS2a)	Inducible	NM 000625.2
eNOS (NOS3)	Constitutive	NM 000603.1

Human nitric oxide synthase enzymes are listed, along with their mode of expression and RefSeq mRNA accession number.

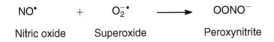

Figure 4.6

Formation of peroxynitrite.

4.3.2 Chemical-mediated signalling

Exposure of humans to larger reactive chemical molecules, either through direct exposure, or as reactive intermediates formed during metabolism elicits a similar range of responses as seen for ROS. A large number of these chemical species effect gene expression through direct interaction with intracellular receptors, which in turn directly activate gene transcription. Direct activation of ligand-activated transcription factors by large chemicals will be discussed in Chapter 7, and will not be further covered here.

As with ROS, indirect activation of gene expression can also be caused by large chemicals, again through the co-ordinated activation of MAPK cascades. The importance of MAPK-mediated signalling in response to xenobiotics may be seen by their obligate role in the cellular responses to peroxisome proliferators. The peroxisome proliferators (PP) are a diverse group of compounds used in a number of industrial, domestic and medical situations. They are negative in all standard genotoxicity assays, yet cause liver cancer in rats and mice during 2-year carcinogenicity studies. PPs elicit a range of changes in rodent liver, including increases in hepatocyte number (hyperplasia) and an increase in the size of individual hepatocytes (hypertrophy), leading to an overall increase in liver size (hepatomegaly).

One feature of PP-mediated hyperplasia is the co-ordinated inhibition of apoptosis and activation of DNA synthesis (Bayly *et al.*, 1994). As shown in *Figure 4.3*, MAPK pathways play a central role in the signal transduction of stimuli that cause apoptosis and DNA synthesis. Pauley *et al.* (2002) examined whether MAPK signalling cascades played a role in peroxisome proliferator-mediated hepatomegaly. They demonstrated that MAPK activation was required for the activation of immediate early genes such as *c-fos* and *c-jun*, two transcription factors involved in the expression of many genes, including those involved in cell cycle control. In contrast MAPK activation appeared to have little effect on peroxisome β-oxidation, the PPARα-mediated response to peroxisome proliferators. One possible conclusion from such studies is that MAPK activation, possibly independent of PPARα activation, is required for the unregulated DNA synthesis/apoptosis that lead to liver growth, and possibly hepatocarcinogenesis as well.

Although Pauley *et al.* suggested that MAPK signal transduction appeared to play only a minor part in the PPARα-mediated response to PPs, they may have some impact on this aspect of the response as well. Misra *et al.* (2002) demonstrated that PBP, a co-activator of PPARs necessary for full transcriptional activation to occur, could be phosphorylated by the MAPK signalling cascade. They engineered HeLa cells which over-expressed Raf-BXB, a constitutively active version of Raf, the MAPKKK required for the eventual activation of the MAPK ERK2 (*Table 4.2*). This led to increased levels of activated ERK2, and a subsequent increase in PBP phosphorylation. This causes an increase in PBP activation, which results in an increase in PPAR-mediated gene expression.

4.4 Repair of cellular damage

As we have seen in the previous section, toxic insult causes a number of cellular changes to occur. Production of reactive intermediates and/or reactive oxygen species leads to a change in the gene expression profile of the cell, designed to combat this potential threat. Up-regulated genes include those encoding for protective enzymes, such as glutathione transferases and superoxide dismutase, with the net effect that chemically reactive species should be removed rapidly before they can interact with cellular components and cause damage. As we will see in the next section, however, such defence systems are not always enough and sometimes the only alternative is the destruction of the cell. In the first scenario cell damage should be zero, with no protein/DNA adducts being formed, whereas in the latter cellular damage is excessive. Where however is the middle ground? Does a situation exist where a certain degree of damage can be tolerated, and indeed recovered from? The answer to this is yes, as amongst the genes activated during the cellular defence programme are those involved in the repair/destruction of damaged proteins/DNA, and it is those we will examine in this section.

4.4.1 Repair of damage to DNA

The body has an array of protective measures to ensure the fidelity of DNA. This is not surprising as DNA is the blueprint of the body; changes in this will affect every downstream process and hence must be avoided. These systems are, in the main, primarily concerned with the fidelity of DNA polymerase during DNA replication. No polymerase is 100% accurate, and a fundamental role of DNA repair enzymes is to 'proofread' replication, replacing any incorrectly copied DNA. However, when the cell is under stress from external sources, such as a chemical insult, these systems, plus others, may be called upon to protect the DNA from external damage, removing and repairing any sections that have been compromised.

Figure 4.7 details the various DNA repair mechanisms within the cell, the type of damage they repair and potential sources of such damage. Mismatch repair is the basic repair system of the cell, removing nucleotides incorrectly incorporated during replication and replacing them with the correct nucleotide. Base excision repair (BER) and nucleotide excision repair (NER) are the main systems to deal with chemical-mediated damage to DNA. Both systems remove one or more damaged nucleotides and replace them with

Mechanism	Mismatch repair	Base excision repair	Nucleotide excision repair	Recombination repair
Damage repaired	A-G or T-C Mismatch	Base modifications, single strand breaks	Large adducts, pyrimidine dimers	Interstrand crosslinks, double strand breaks
Potential source of damage	Normal DNA Replication	ROS, alkylating agents, X-rays	Reactive intermediates, UV radiation	Clastogens, X-rays

Figure 4.7

DNA repair mechanisms (adapted from Hartwig and Schwerdtle, 2002).

the correct nucleotide; BER is targeted towards small adducts or DNA lesions (e.g. ROS-mediated damage) whereas NER is a more flexible system that can remove larger adducts, such as those formed by interactions of reactive chemical intermediates with DNA. Finally, large chromosomal aberrations, such as those caused by clastogens, have their damage repaired by recombination repair systems.

One interesting aspect of toxic damage is that the DNA repair mechanisms themselves may be targets for toxic chemicals. For example, metals such as nickel and cobalt are carcinogenic, but their mutagenic potentials are actually rather poor. The way they increase their carcinogenic effect is to act as co-mutagens, damaging DNA repair mechanisms, specifically base excision repair and nucleotide excision repair systems, although the exact mode by which this is achieved is unknown (Hartwig and Schwerdtle, 2002). Hence, their action is twofold; firstly to cause a low level of DNA damage through their weak mutagenic potential, and secondly to prevent the cell recovering from this DNA damage.

4.4.2 Repair of damage to proteins

As with repair of DNA damage, the enzymes involved in repair of protein damage play normal physiological roles; protein degradation is an important part in the regulation of protein levels within the cell and hence in controlling their actions. The best understood protein degradation pathway is the ubiquitin pathway, which targets protein for degradation by the proteosome. Ubiquitin is a small (76 amino acid) protein that, once conjugated to lysine residues of proteins, targets them for destruction. The addition of a single ubiquitin molecule however provides only a weak signal for degradation; it is the attachment of polyubiquitin chains that commit a protein to degradation. This conjugation is catalysed by three enzymes, termed ubiquitin-activating enzyme (E1), ubiquitin-conjugating enzyme (E2) and ubiquitin-protein ligase (E3), which sequentially add successive ubiquitin monomers to form the polyubiquitinated protein (*Figure 4.8*). This protein is then targeted by the 20S proteosome, a barrel-like structure of 14 α and 14 β subunits.

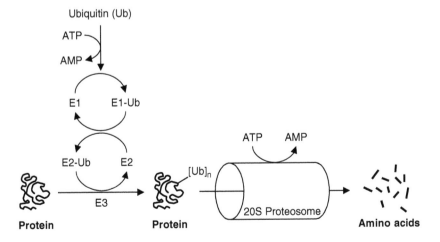

Figure 4.8

The ubiquitin system of protein degradation.

The ubiquitinated protein passes through the centre of the 20S proteosome where it is sequentially cleaved into its component amino acids (*Figure 4.8*).

So how is the degradation of proteins by ubiquitination linked to cellular defence to chemical insult? Changes in the levels of ubiquitination enzymes (E1, E2 and E3) are often associated with exposure to toxic chemicals. In addition, proteins damaged by adduct conjugation have been shown to be targets for ubiquitination (see Donohue (2002) for review). Hence, ubiquitination plays a vital role in the cellular defence against chemical insult, removing damaged proteins so that they can be replaced by fully functioning counterparts.

4.5 Regulation of apoptosis and necrosis

While there exists within cells protective mechanisms to recover from a toxic insult, such as those described above, there will come a point beyond which the cell can no longer be saved. When cellular damage reaches these levels two possible cell fates exist, destruction via necrosis or destruction via apoptosis. These two processes may have the same end point but their mode of functioning and biological consequences are very different.

4.5.1 Necrosis

Necrosis is defined as the death of an area of tissue, usually surrounded by healthy tissue (Hodgson *et al.*, 1998). It is associated with a number of biological processes, most notably inflammation, lysosomal enzyme release and cellular fragmentation. Necrosis is often referred to as an 'uncontrolled' form of cell death, as opposed to the 'programmed' cell death of apoptosis. Such an assumption, while generally true, does hide some important facts about necrosis. Firstly, the processes are not truly 'uncontrolled' as many of the biochemical changes associated with necrosis are mediated by enzymes, and therefore, as with all enzymatic reactions, have the potential to be controlled. In addition, many of the key characteristics of apoptosis are also seen during necrosis, including nuclear disintegration (karyorrhexis) and loss of nuclear material (karyolysis; Kanduc *et al.*, 2002). Such similarities are not surprising when you consider that it is now becoming clear that many cases of cell death embrace both of these processes, with apoptosis occurring initially, swapping to necrosis as toxic injury continues (Pierce *et al.*, 2002). Hence, necrosis is probably more correctly referred to as the changes that occur later in tissue damage, whereas apoptosis, as we will see, is the controlled process that leads up to this point.

4.5.2 Apoptosis

In direct contrast to the uncontrolled nature of necrotic cell death, apoptosis may be defined as programmed cell death, characterized by high levels of energy consumption, condensation of nuclear chromatin, internucleosomal cleavage of DNA, and activation of highly conserved signal transduction pathways (Hodgson *et al.*, 1998). The presence of a programmed form of cellular destruction was postulated for many years, but it was not until 1972 that it was described and given the name apoptosis (Kerr *et al.*, 1972). In the following three decades, apoptosis has been shown to be a key regulator in many biological functions, from organogenesis to response to toxins. Many of the molecular mechanisms of apoptosis have been described in the past decade, although the complete picture is still not fully known. What signals a cell to undergo apoptosis and how does this process occur?

Two sets of proteins appear to play key roles in apoptosis; the caspases and Bcl-2 family members. Caspases are a family of cysteine proteases that, once activated, cleave proteins at aspartic acid residues. In man, 12 caspases have so far been identified (*Table 4.3*), and the specificity of each towards its target proteins is set by the three amino acids directly following the aspartic acid residue. This cleavage can result in either activation or, more usually, inactivation of the target protein and it is this series of cleavages that produces the signalling cascade that ultimately results in apoptosis.

If caspases are the central activators of the apoptosis cascade how then are they initially activated? Three potential routes of activation exist.

1. Aggregation of caspase zymogens: While the zymogen molecule is often described as inert, it does in fact possess very low levels of caspase activity. Upon activation, apoptosis-inducing receptors such as CD95 cause aggregation of caspase-8 zymogen molecules at the membrane. It has been hypothesized that this close proximity allows the low-level activity to cleave the zymogens and produce active caspase-8, which can then cleave other caspase-8 zymogen molecules, so setting up an amplification loop. Such a process may explain how the initial step in the caspase-mediated cascade, activation of caspase-8, is achieved.
2. Cleavage by a caspase: In common with most proteases, caspases are stored within the cell as inactive precursors (zymogen molecules). They hence require cleavage to remove the prodomain and form the active protein. Many caspases are activated by cleavage mediated by other caspases further upstream in the signalling cascade.
3. Association with other proteins: Cleavage of the caspase-9 zymogen molecule produces little increase in the cleavage activity of this caspase. Instead, activity is regulated by association with Apaf-1, a co-factor that is required for correct functioning of this enzyme. In addition to association

Table 4.3   Human caspases

Name	Cleavage targets	Refseq Accession
Caspase-1	IL-1β	5 alternate transcripts
Caspase-2	Unknown	4 alternate transcripts
Caspase-3	Caspase-6, -7, -9	2 alternate transcripts
Caspase-4	Caspase-1	3 alternate transcripts
Caspase-5	Unknown	NM_004347.1
Caspase-6	PARP	2 alternate transcripts
Caspase-7	PARP	4 alternate transcripts
Caspase-8	Caspase-3, -4, -6, -7, -8, -9, -10, PARP	5 alternate transcripts
Caspase-9	Various (see Table 4.5)	2 alternate transcripts
Caspase-10	Caspase-3, -4, -6, -7, -8, -9	4 alternate transcripts
Caspase-13	Unknown	Unknown
Caspase-14	Unknown	NM_012114.1

Human caspases are listed, along with their major intracellular cleavage targets, and RefSeq mRNA accession number or number of alternate transcripts. (a) Cleavage of IL-1b by caspase-1 produces the active version of this cytokine.

with proteins that act in a positive fashion, a number of proteins have been identified that negatively affect caspase activity, including the inhibitors of apoptosis (IAPs). These proteins themselves are also regulated by other proteins, such as Smac, a protein that blocks the inhibitory action of IAPs.

A representation of the putative activation cascade for apoptosis is given in *Figure 4.9*, and comprehensively reviewed by Hengartner (2000). The major effect of pro-apoptotic stimuli is the sequential activation of caspases. Initially, caspase-8 is activated by cleavage from its inactive zymogen (point 1 above) and this then causes the sequential activation of other caspases through their cleavage (point 2). Finally caspase-9 associates with Apaf1 and cytochrome c to form the active apoptosome (point 3), which targets many cellular proteins for cleavage and commits the cell to an apoptotic death.

While the central cascade for apoptosis appears to be activation of caspases, there are several other protein families that play important roles in modulating this cascade, perhaps the most important being the Bcl-2 family, a group which possesses both anti-apoptotic and pro-apoptotic effects (*Table 4.4*). Bcl-2 family members are capable of interacting with other members of the family, forming heterodimers that cancel out the effects of each partner, or homodimers, which results in an active protein. Hence, on a simplistic level,

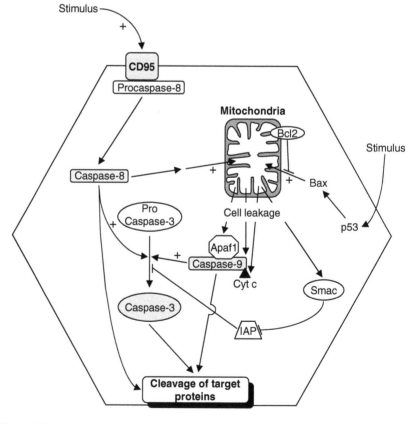

Figure 4.9

Overview of apoptosis (adapted from Hengartner, 2000).

Table 4.4   Human Bcl-2 family members

Pro-apoptotic members		Anti-apoptotic members	
Name	RefSeq	Name	RefSeq
Bax	6 alternate transcripts	Bcl-2	2 alternate transcripts
Bcl2L7 (Bak1)	NM 001188.1	Bcl2L1 (BclX)	2 alternate transcripts
Bcl2L8 (Bad)	2 alternate transcripts	Bcl2L2 (BclW)	NM 004050.2
Bcl2L9 (BokL)	2 alternate transcripts	Bcl2L5 (BFL1)	NM 004049.2
Bcl2L11	8 alternate transcripts	Bcl2L10 (BclB)	NM 020396.2
Bcl2L12	2 alternate transcripts	BNIP 1	4 alternate transcripts
Bcl2L13 (Rambo)	NM 015367.1	BNIP2	NM 004330.1
BNIP3	NM 004052.2	Mcl1	NM 021960.2
Bid	NM 001196.1	BAG1	NM 004323.2
Bid3 (Hrk)	NM 003806.1	BAG2	NM 004282.2
Bik	NM 001197.3	BAG3	NM 004281.2
		BAG4	NM 004874.1
		BAG5	NM 004873.1

Human Bcl-2 family members are listed, with common alternate names in brackets and RefSeq mRNA accession number or number of alternate transcripts.

if a cell contains more pro-apoptotic Bcl-2 family members than anti-apoptotic ones it will form more pro-apoptotic homodimers and thus be predisposed to undergo apoptosis. For example, as can be seen from *Figure 4.9*, Bax, a pro-apoptotic member of the Bcl-2 family that is activated by p53-mediated stimuli, can have its action attenuated by the anti-apoptotic activity of Bcl-2. If the balance within a cell is such that there is more pro-apoptotic Bax than anti-apoptotic Bcl-2, then Bax will stimulate disruption of the mitochondrial membrane, allowing several key proteins to leak out, although the exact mechanism by which this occurs is not certain. Among the proteins that leak out of mitochondria are pro-caspases, further amplifying the activation cascade, Smac and cytochrome c. The last of these, cytochrome c, is usually associated with the electron transport chain and energy production within the cell interacts with caspase-9 and Apaf-1 to form the apoptosome, a complex that causes many of the downstream effects of apoptosis (*Figure 4.9*).

Following activation of the caspase cascade what then is the biological result? As detailed above caspases are proteolytic enzymes and hence their biological role is to cleave proteins, either activating of deactivating their biological functioning. At present over 100 intracellular targets for caspase cleavage have been identified (Earnshaw *et al.*, 1999). Although many of these may just be 'caught in the cross-fire' of the cleavage cascade, a reasonable number would be expected to be involved in the morphological characteristics of apoptosis, including nuclear condensation, DNA cleavage, cell shrinkage and phagocytosis. *Table 4.5* lists some of the classes of proteins targeted for cleavage by caspases and examples of proteins that belong to each of these classes.

As our basic understanding of 'programmed cell death' increases an attractive proposition is the utilization of this in medicine: could we specifically target, for example, cancer cells for destruction? One such application is the

Table 4.5 Cleavage targets for mammalian caspases

Protein class	Example of proteins within class
Established role in apoptosis	Bcl-2, Bcl2L1, Bax, ICAD
Abundant cytosolic proteins	Actin, gelsolin, α-fodrin
Abundant nuclear proteins	Lamin A and B, hnRNP proteins C1 and C2
DNA metabolism and repair	PARP, RAD51, DNA topoisomerase II
Protein kinases	PKC-β1, δ, and Θ, p21-activated kinase
Signal transduction (not kinases)	Pro-interleukin 1β, 16 and 18, NF-κB
Cell cycle and proliferation	P21Cip1/Waf1, p27Kip1, CDC27
Role in human genetic disease	Huntingtin, presenilin-1 and 2

Data from Earnshaw *et al.* (1999).

use of Bcl-2 antisense technology. The anti-apoptotic protein Bcl-2 was first identified in B-cell lymphomas, and work with transgenic mice has confirmed its role in tumour development – unrestrained cell proliferation is exacerbated by diminished cell removal through apoptosis (Veis *et al.*, 1993). A novel anticancer approach would therefore be to disrupt Bcl-2 activity, thus increasing the rate of apoptosis in the tumour and slowing its progression. Genta, an antisense oligomer that binds to the Bcl-2 mRNA and prevents its translation into protein, is the first, successful, attempt to exploit such a target. Human tumours xeno-transplanted into mice treated with Genta showed greater than 90% decrease in their rate of growth compared to controls, which is better than the response seen with traditional chemotherapeutic agents such as cisplatin, showing the potential of such antisense technology. In addition, co-administration with standard chemotherapeutic agents showed an increased efficacy, with the lower levels of Bcl-2 caused by Genta treatment making the tumours more chemosensitive (Schlagbauer-Wadl *et al.*, 2000). Genta is currently undergoing Phase II clinical trials.

4.6 Summary

In this chapter we have examined how the body co-ordinates its overall response to toxic insult. Chapters 2 and 3 detailed the individual reactions that occur within the body to facilitate excretion of chemicals from the body, namely Phase I and II metabolism. We also saw how an imbalance of these two phases could lead to a build-up of reactive chemicals, and how this resulted in toxicity. However, it is important to remember that not all toxicity is metabolism-derived. Many compounds that we are exposed to are inherently toxic and do not require bioactivation. This chapter has examined how the body responds to toxic insult regardless of source. We have seen that following stimuli a series of signalling pathways is used to co-ordinate the changes in gene expression required to express the proteins that will combat the toxic insult. These proteins may take one of two forms. In the first case they are designed to remove the toxic chemical before serious cellular damage can occur; this is obviously the first-choice system. However, if damage is severe then other pathways can be activated that lead to cellular death, removing compromised cells before they can do any damage to the entire organ.

In Chapter 5, we will look at how this co-ordinated response is utilized to respond to some specific examples of toxicity, each affecting a different facet of the body's normal functioning.

References

Bayly, A.C., Roberts, R.A. and Dive, C. (1994) Supression of liver cell apoptosis *in vitro* by the non-genotoxic hepatocarcinogen and peroxisome proliferator nafenopin. *J. Cell Biol.* **125**(1): 197–203.

Beck, K.F., Eberhardt, W., Frank, S., *et al.* (1999) Inducible NO synthase: role in cellular signalling. *J. Exp. Biol.* **202**: 645–653.

Chuang, Y.Y., Chen, Y., Gadisetti, T., *et al.* (2002) Gene expression after treatment with hydrogen peroxide, menadione, or t-butyl hydroperoxide in breast cancer cells. *Cancer Res.* **62**(21): 6246–6254.

Donohue, T.M., Jr. (2002) The ubiquitin-proteasome system and its role in ethanol-induced disorders. *Addiction Biol.* **7**(1): 15–28.

Earnshaw, W.C., Martins, L.M. and Kaufmann, S.H. (1999) Mammalian caspases: structure, activation, substrates, and functions during apoptosis. *Ann. Rev. Biochem.* **68**: 383–424.

Guyton, K.Z., Liu, Y., Gorospe, M., *et al.* (1996) Activation of mitogen-activated protein kinase by H_2O_2. Role in cell survival following oxidant injury. *J. Biol. Chem.* **271**(8): 4138–4142.

Hartwig, A. and Schwerdtle, T. (2002) Interactions by carcinogenic metal compounds with DNA repair processes: toxicological implications. *Toxicol. Lett.* **127**(1–3): 47–54.

Hengartner, M.O. (2000) The biochemistry of apoptosis. *Nature* **407**(6805): 770–776.

Hodgson, E., Mailman, R. and Chambers, E. (1998) *Dictionary of Toxicology*, 2nd edn. Macmillan Reference Ltd., London.

Itoh, K., Wakabayashi, N., Yasutake, *et al.* (1999) Keap1 represses nuclear activation of antioxidant responsive elements by Nrf2 through binding to the amino-terminal Neh2 domain. *Genes Develop.* **13**: 76–86.

Kanduc, D.K., Mittleman, A., Serpico, R., *et al.* (2002) Cell death: apoptosis versus necrosis. *Int. J. Oncol.* **21**: 165–170.

Kerr, J.F., Wyllie, A.H. and Currie, A.R. (1972) Apoptosis: a basic biological phenomenon with wide-ranging implications in tissue kinetics. *Br. J. Cancer* **26**: 239–257.

Kim, D.H., Feinbaum, R., Alloing G., *et al.* (2002) A conserved p38 MAP kinase pathway in *Caenorhabditis elegans* innate immunity. *Science* **297**: 623–626.

Liebermann, D.A., Hoffman, B. and Steinman, R.A. (1995) Molecular controls of growth arrest and apoptosis: p53-dependent and independent pathways. *Oncogene* **11**: 199–210.

Misra, P., Owuor, E.D., Li, W., *et al.* (2002) Phosphorylation of transcriptional coactivator PBP: stimulation of transcriptional regulation by mitogen-activated protein kinase. *J. Biol. Chem.* **277**(50): 48745–48754.

Morel, Y. and Barouki, R. (1999) Repression of gene expression by oxidative stress. *Biochem. J.* **342**(3): 481–496.

Pauley, C.J., Ledwith, B.J. and Kaplanski, C. (2002) Peroxisome proliferators activate growth regulatory pathways largely via peroxisome proliferator-activated receptor alpha-independent mechanisms. *Cellular Signalling* **14**(4): 351–358.

Pierce, R.H., Franklin, C.C., Campbell, J.S., *et al.* (2002) Cell culture model for acetaminophen-induced hepatocyte death in vivo. *Biochem. Pharmacol.* **64**(3): 413–424.

Ryan, T.P. and Aust, S.D. (1992) The role of iron in oxygen-mediated toxicities. *Crit. Rev. Toxicol.* **22**: 119–141.

Sakurai, K. and Cederbaum, A.I. (1998) Oxidative stress and cytotoxicity induced by ferric-nitriloacetate in HepG2 cells that express Cytochrome P450 2E1. *Mol. Pharmacol.* **54**: 1024–1035.

Schaeffer, H.J. and Weber, M.J. (1999) Mitogen-activated protein kinases: specific messages from ubiquitous messengers. *Mol. Cell. Biol.* **19**(4): 2435–2444.

Schlagbauer-Wadl, H., Klosner, G., Heere-Ress, E., *et al.* (2000) Bcl-2 antisense oligonucleotides (G3139) inhibit Merkel cell carcinoma growth in SCID mice. *J. Investigative Dermatol.* **114**: 725–730.

Sumbayev, V., Sandau, K. and Brune, B. (2002) Mesangial cells but not hepatocytes are protected against NO/O(2)(−) cogeneration: mechanistic considerations. *Eur. J. Pharmacol.* **444**(1–2): 1.

Veis, D.J., Sorensen, C.M., Shutter, J.R., *et al.* (1993) Bcl-2-deficient mice demonstrate fulminant lymphoid apoptosis, polycystic kidneys and hypopigmented hair. *Cell* **75**: 229–240.

Toxicity case studies

5

In this chapter we will study a number of examples of toxicity and examine how the chemicals involved cause their effects. We will see how the site of toxicity is determined by a number of factors, including route of exposure, distribution within the body and sites of metabolism. In addition we will examine how the use of molecular tools has allowed the further delineation of molecular mechanisms underlying these toxicities. Such advances have important ramifications for risk assessment and treatment of exposure as well as in the design of novel compounds that lack the toxicological drawbacks but maintain the therapeutic benefits of their parent compounds.

5.1 Genotoxicity

Genotoxicity may be defined as a 'specific adverse effect on the genome of living cells that, upon the duplication of the affected cells, can be expressed as a mutagenic or a carcinogenic event' (Hodgson *et al.*, 1998). As we have seen in previous chapters, many compounds produce chemically reactive species during metabolism, or are themselves reactive and may therefore cause changes that fit into this category. Hence, the field of genotoxicity requires further sub-divisions to allow easy classification and quantification of these effects. One of the simplest divisions is based upon the specific type of interaction that the compound has with the genome, and the resultant effect that it has upon the genome. Using such an operational distinction we can divide genotoxicity into three broad categories. Aneugens induce unequal segregation of chromosomes during cell division and result in a cell with more or fewer chromosomes than normal (aneuploidy). Clastogens also affect whole chromosomes, but rather than effecting the segregation of chromosomes during meiosis they cause gross structural changes, primarily through DNA cleavage. Finally, mutagens are chemicals which do not affect DNA at the level of the chromosome; instead localized changes in the sequence of DNA are induced by this class of chemical.

Such definitions are important not only to simplify the field, but also in terms of risk assessment. Aneugens very often exhibit a threshold concentration, below which no genotoxic effects are seen. In terms of risk assessment, exposure limits for aneugens may therefore be set according to the experimentally determined 'no effect level'. In comparison, chemicals that damage DNA directly (i.e. clastogens and mutagens) are thought of as having no threshold for effect; i.e. any concentration of compound will result in some DNA damage. How do you set a 'safe' exposure limit? In general clastogens and mutagens have their exposure limits set according to a linear dose–response curve for the observed effects at higher doses, and this tends to result in much lower 'safe exposure' levels than a threshold-based calculation.

Examples of each of these classes will be discussed below.

5.1.1 Aneugens: vinca alkaloids

As defined above, aneugens are chemicals that cause abnormal numbers of chromosomes within a cell. This is achieved through uneven chromosome segregation during cell division, resulting in one daughter cell with more than the normal complement of chromosomes and one with less. The mode of action of these compounds is generally through disruption of mitotic spindle formation, the structures involved in separation of chromosomes during meiosis. During meiosis two spindles are formed and migrate to the opposite poles of the dividing cell. Microtubules, formed of polymers of α- and β-tubulin, radiate out from each spindle and attach to the sister chromatids formed during DNA replication. These newly generated chromosomes are then drawn apart to the opposite poles of the cell, meaning that one copy of each chromatid segregates to each of the resulting daughter cells. When spindle formation is disrupted the chromosomes segregate randomly, resulting in uneven numbers of chromosomes in each daughter cell. This chromosomal imbalance is often lethal to the cell and hence many aneugens are highly cytotoxic. Due to the fact that aneugens also only target actively dividing cells, many of these compounds have found a clinical use in the treatment of cancer, including the vinca alkaloids.

Vincristine and vinblastine are two naturally occurring alkaloids derived from the *Catharanthus roseus* plant (*Figure 5.1*). As described above, spindles and their tubulin radiations are generated from the concatenation of α/β tubulin heterodimers, and it is to these α/β heterodimers that the vinca alkaloids bind. In doing so they block the GTP-binding site on the heterodimers and this prevents the addition of further dimers to the tubule, thus preventing complete spindle molecule formation and function. Both compounds have widespread use in chemotherapy, although with differing target cancers (*Table 5.1*).

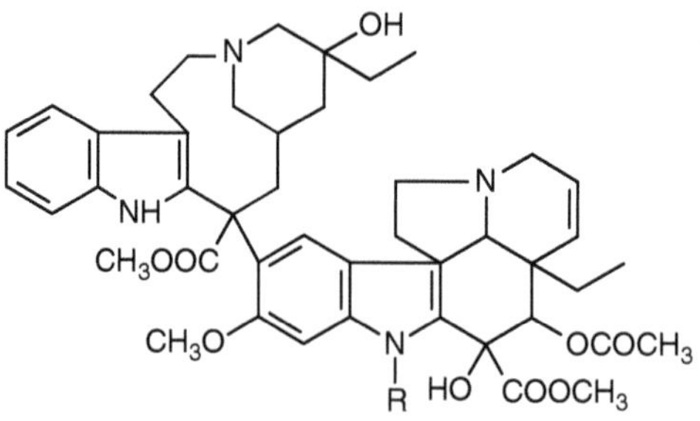

Vincristine: R＝CH₃
Vinblastine: R＝CHO

Figure 5.1

Structure of vinca alkaloids.

Table 5.1 Use of vinca alkaloids in treating cancer

Vincristine	Vinblastine
Acute leukaemia	Lymphocytic leukaemia
Neuroblastoma	Histiocytic leukaemia
Rhabdomyosarcoma	Karposi's sarcoma
Hodgkin's lymphoma	Hodgkin's lymphoma
General lymphomas	Advanced breast cancer
	Advanced testicular cancer

One problem associated with cancer chemotherapy is that of induced resistance. We already know that the expression of enzymes involved in drug metabolism may be increased in response to drug exposure, thus quickly and efficiently removing the compound from the body. Unfortunately such effects also exist for compounds that we wish to keep in the body, such as toxic chemotherapeutic drugs. In this case, much of this resistance is due to the induction of membrane transporter proteins such as P-glycoprotein (Pgp) and multidrug-resistance-related proteins (MRP1-5), which pump drugs away from the target cells and therefore decrease their efficacy. Indeed, drugs targeted to disrupt these membrane transporters are currently available, although the associated adverse side effects are severe. To increase the efficacy of anticancer drugs it is therefore important to delineate other mechanisms of resistance, some of which may be more open to therapeutic blockade than the transport molecules. Wang *et al.* (2002) used differential display technology to examine the changes in gene expression seen *in vitro* following exposure to vincristine. They identified 20 mRNA species that were expressed at high level following treatment with vincristine. While seven of these corresponded to known mRNA species, of which some, including drug transporters would be hypothesized to impact upon multidrug resistance, the remainder were novel sequences and thus provide new targets for investigation. Thus, using this technology it was possible to identify several new lines of enquiry for the molecular mechanisms underlying multidrug resistance and may provide new therapeutic targets to increase the efficacy of anticancer chemotherapy.

As stated in the introduction, aneugens are subject to threshold-based risk assessment whereas clastogens and mutagens are not, due to the lack of a true no-effect level. However, accumulating evidence suggests that the division between aneugens and clastogens is not always distinct. Chromosomal aberrations have been reported for a number of classical aneugens, such as vincristine, and this therefore raises the question of whether these compounds can be subject to a threshold-based risk assessment or not. Examination of the types of chromosomal aberrations that occur suggest that they are probably as a result of the aneugenic properties of the compounds (i.e. disruption of the spindle and/or their attachment to chromosomes) and not due to direct DNA-damaging effects (Arni and Hertner, 1997). Hence, the current threshold-based risk assessment does appear to be appropriate.

5.1.2 Clastogens: cadmium chloride

Compounds that cause aneuploidy do not directly interfere with DNA, but rather exert their effects by altering the segregation of DNA during meiosis. Clastogens cause chromosome alterations, but achieve this via acting directly on DNA, inducing strand breaks and displacements.

Cadmium is one of a number of heavy metals with widespread industrial usage, including roles in the production of alloys, metal plating, and in the manufacture of a variety of pigments. Such use leads to a potential occupational exposure, through inhalation, during any of these processes. Cadmium is also present within the diet, but due to poor intestinal absorption this is not a major route of exposure.

Cadmium toxicity manifests as liver and testicular damage following acute exposure whereas chronic exposure leads to kidney damage. The exact cause of this damage is unknown, although recent data suggest that the generation of reactive oxygen species by cadmium causes the observed clastogenic effects.

The role of metallothionein proteins (MTs) in the cellular response to cadmium has been investigated by several groups. MTs are a major group of intracellular proteins which bind to zinc and exhibit antioxidant properties. Lazo *et al.* (1995) engineered transgenic mice lacking the genes encoding for MTI and MTII and then examined their sensitivity to cadmium – the mice were considerably more sensitive to its toxic effects. This effect is hypothesized to be due to the role of MTs in apoptosis: MTs act as a stress sensor, targeting cells for apoptosis in response to toxic stimuli, including the DNA damage caused by cadmium. Removal of the MTs therefore prevents the undergoing apoptosis and they survive, complete with the induced chromosomal aberrations.

Using DNA microarray technology, Joseph *et al.* (2002) examined the response of cells to cadmium exposure. They identified a novel translation elongation factor (1δ), which was consistently up-regulated by cadmium exposure. In addition, blocking this up-regulation using antisense mRNA prevented the oncogenic effect of cadmium, strongly suggesting the up-regulation of this protein is a cause of the oncogenic phenomenon and not a consequence. Joseph *et al.* hypothesized that over-production of this elongation factor may result in increased translation and, in turn, the loss of proofreading ability. Such translational infidelity could result in production of malfunctioning proteins that, in combination with the clastogenic effects, greatly increase the oncogenic potential of cadmium.

5.1.3 Mutagens: vinyl chloride

The third class of genotoxic chemicals are those that interact with DNA on a smaller scale, although the end result can be just as toxic. Mutagens interact with only small regions of DNA and cause localized damage to the double helix.

Vinyl chloride is the monomeric component of the widely used plastic polyvinyl chloride (PVC). During the manufacture of PVC workers may be exposed to monomeric vinyl chloride, and acute exposure has historically been associated with a number of adverse effects, including vertigo, lethargy, hearing and vision loss and even, following high-dose exposure, loss of consciousness. Chronic exposure results in a relatively rare form of liver cancer, haemangiosarcoma. Work in experimental animals demonstrated that this was due to cytochrome P450-mediated metabolism to form a reactive epoxide, which, unless it is detoxified by either glutathione conjugation or the action of epoxide hydrolase, can react with DNA and/or proteins to produce adducts (*Figure 5.2*).

As described in the introduction, mutagens do not have a 'no-effect level' for the purposes of risk assessment. If we state that the presence of a mutagen *will* result in the formation of mutations how then can we set a level at which exposure is 'safe'? Through the use of high-sensitivity immunoaffinity/GC-high resolution MS it is now clear that the formation of endogenous DNA adducts is a relatively common phenomenon and we may therefore rephrase our initial question (La and Swenberg 1996). Instead of stating that the presence of *any* DNA adduct is bad, we must now ask at what level of xenobiotic-induced adducts are the DNA repair mechanisms (Chapter 4) placed under significant extra load, thus significantly increasing the risk to

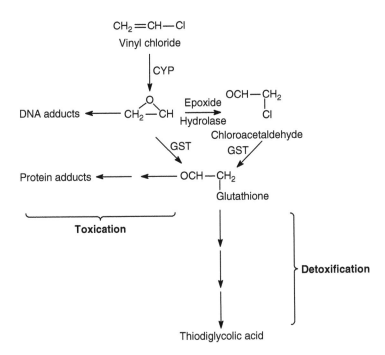

Figure 5.2

Toxicity of vinyl chloride.

the individual; based on this we can then set this level as our 'threshold of effect'. Swenberg *et al.* (2000) examined the effect of low level exposure of rats to vinyl chloride on DNA adducts, in particular the *N*2,3-ethenoguanine which has been implicated in miscoding during DNA synthesis, and result-ant carcinogenesis. They demonstrated that while high level exposure (100 ppm) resulted in a 25-fold increase in the level of *N*2,3-ethenoguanine adducts, lower exposure (10 ppm) resulted in only a 5.9-fold increase. This large difference in the formation of adducts between the two doses may go some way to explaining why no cases of haemangiosarcoma have been reported since exposure limits have been reduced to 1 ppm; prior to this it was estimated that occupational exposure was greater than 100 ppm. There-fore, with the advent of more sensitive techniques we have been able to rephrase our question for risk assessment, and go some way to setting a 'no *significant* effect level' for mutagens.

5.2 Hepatotoxicity

The liver, as the major organ responsible for metabolism within the body, is the site where many compounds exert toxic effects. As the liver is therefore often the primary target for toxicity, if we wish to search for markers of toxicity the liver is a good candidate organ to begin studying. However, it is of little long-term value to identify markers specific for a single chemical, instead we must identify markers that are chemical-class or toxic-endpoint specific. Such markers allow the screening of new compounds to identify possible toxicities associated with them, and group them together with known, well-studied, toxins that cause the same adverse effects; this approach would aid in the identification of the possible mechanisms underlying the observed toxicity.

In an attempt to achieve such an aim, much research using microarray analysis has been undertaken to examine the changes in gene expression observed in the liver following exposure to classical hepatotoxins. In one example, Waring *et al.* (2001) examined the changes in gene expression profile in rat liver following exposure to 15 known hepatotoxins. Rats were exposed to each of the compounds for 3 days, and at doses previously shown to cause severe hepatoxicity after 7 days of dosing. Using this regime it was hoped to identify gene expression changes that occurred *before* histological changes were evident, thus increasing their predictive value. Using an Affymetrix GeneChip test2 microarray Waring *et al.* identified changes in gene expression resulting from toxicant exposure and then used three separate clustering methods to distinguish clusters of compounds that caused statistically similar changes in gene expression profiles. Using this approach it was possible to cluster compounds in terms of their effects on gene expression, and this clustering showed correlation with clusters derived from both the histopathology and clinical biochemistry data. For example, the gene expression profiles clustered Aroclor 1254 and 3-methylcholanthrene together, and both of these compounds are known to cause liver hypertrophy through increases in smooth endoplasmic reticulum. One interesting result was the clustering of carbon tetrachloride and allyl alcohol. While both of these compounds cause liver necrosis, carbon tetrachloride damage is localized to the centrilobular region while allyl alcohol damages the periportal region, suggesting different mechanisms. However, both compounds cause their effects through radical formation. It is thus possible that the clustering reflects a central mechanism of liver response to damaging radicals (i.e. haem oxygenase, glutathione transferase), rather than the idiosyncrasies of each individual toxin. Such an approach therefore provides the ability to identify marker genes, for which changes in expression profile are markers of potential adverse effects, and that can be used early in the discovery phase of compound development. This will aid in the selection of likely candidate compounds to progress to more in-depth studies.

5.2.1 Carbon tetrachloride

Carbon tetrachloride is a classical liver toxicant, causing massive centrilobular necrosis in a number of species. While it does have extrahepatic effects, these are secondary to the liver necrosis, tend to be sex- and species-dependent and hence will not be discussed herein. Carbon tetrachloride hepatoxicity is dependent upon metabolic activation by CYP2E1 and CYP2B1 in the liver to produce a trichloromethyl radical (*Figure 5.3*). This reaction is interesting for two reasons; firstly the reaction itself is not a characteristic CYP-mediated oxidation reaction but instead an example of CYP-mediated reductive metabolism. Secondly, although the trichloromethyl radical is in itself chemically reactive, it is not the major agent that causes liver damage. Instead the radical interacts with polyunsaturated lipids to produce chloroform and a lipid radical, the latter of which causes a chain reaction of lipid peroxidation reactions and is responsible for the majority of cellular damage associated with carbon tetrachloride exposure.

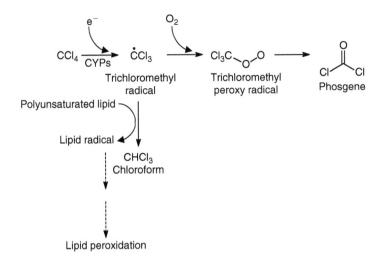

Figure 5.3

Toxicity of carbon tetrachloride.

Another aspect of carbon tetrachloride toxicity is the promotion of fibrosis, the deposition of collagen associated with the repair phase of the inflammatory response. Fibrosis has been suggested as a generalized response to hepatotoxins, caused through the activation of cytokines. Simeonova *et al.* (2001) examined the role of TNFα, a pro-inflammatory cytokine, in this process, as both fibroblast proliferation and collagen accumulation are known to be influenced by pro-inflammatory cytokines. They exposed mice deficient in either a single TNF-receptor, TNFR1 (−/−) or TNFR2 (−/−), or double knockout mice (TNF-DKO) to a toxic dose of carbon tetrachloride and then compared the histological changes following acute or chronic exposure. The immediate, acute response, caused by the lipid peroxidation described above, was unaffected in the knockout animals. This is not unexpected as TNFα has never been implicated in this phase of the toxicity. However, fibrosis was markedly reduced in TNFR2 (−/−) mice and almost absent in TNF-DKO mice, strongly supporting the role of TNFα in this response.

Therefore, it can be seen that carbon tetrachloride produces a two-stage toxic response. The first is an acute, radical-mediated necrotic injury, whereas the second phase is a chronic inflammatory response resulting in liver fibrosis.

5.3   Nephrotoxicity

In a similar fashion to the liver, the kidney is an organ particularly susceptible to the toxic effects of compounds. The underlying reasons for the apparent targeting of these two organs are also similar. Both receive large amounts of blood flow (with the kidneys receiving 25% of cardiac output), meaning that compound distribution to these organs is high. In addition, both have high metabolic capacities and may thus metabolically activate compounds. A further reason for the extent of toxicity observed in the kidneys is the pivotal role this organ plays in excretion. The kidneys are one of the major organs involved in excretion of small metabolites (Mr < 300), and due to this these compounds may accumulate in the kidney, resulting in levels of potentially toxic compounds in the kidney that are higher than in other organs.

5.3.1 Antibiotics

Aminoglycosides, such as streptomycin, are a large group of antibiotics composed of amino sugars joined by a glycosidic linkage to inositol (*Figure 5.4*). Aminoglycosides are used in the treatment of Gram-positive bacterial infections and act by inhibiting translation at the step of translocation of the nascent protein chain from the A to P sites of the ribosome complex (*Figure 5.5*). Under normal use aminoglycoside-mediated toxicity is not a major concern as the levels of antibiotic required to inhibit bacterial translation are not sufficiently high to adversely affect translation in the cells of

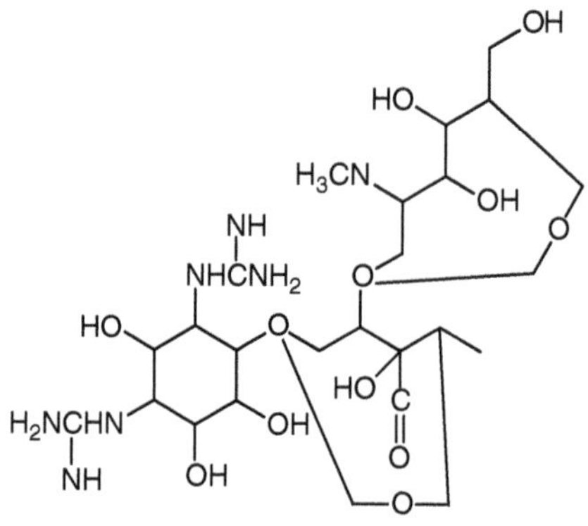

Figure 5.4

Structure of the aminoglycoside Streptomycin.

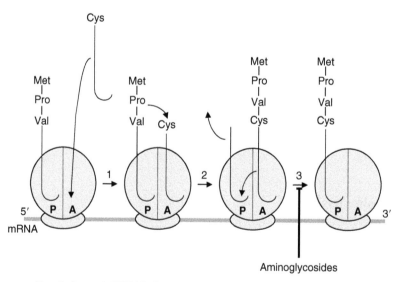

(1 = Aminoacyl-tRNA binding; 2 = Transpeptidation; 3 = Translocation)

Figure 5.5

Disruption of translation by aminoglycosides.

the infected individual. However, due to the aforementioned concentration effect of chemical in the kidney, levels of aminoglycosides can reach five times that seen in the blood, and 5–10% of individuals taking aminoglycoside antibiotics suffer some degree of nephrotoxicity.

Aubrecht *et al.* (1997) examined the possibility of gene therapy as a tool to counteract such toxic effects. They engineered transgenic mice that expressed the hygromycin B phosphotransferase gene (hyg^R), which breaks down the aminoglycoside hygromycin, and examined the effect on nephrotoxic doses of hygromycin on kidney function. Transgenic mice were much more resistant to the toxic effects of hygromycin, with the lethal dose increasing nearly 90-fold compared to that seen in wild-type mice. If such technology could be targeted to produce expression of hyg^R in the kidney only, such technology could present a breakthrough in the prevention of the major source of adverse side effects associated with aminoglycoside antibiotic treatment. While such an approach is rather drastic, and more careful monitoring of the clearance of antibiotic in the urine could also prevent potential toxicity, this does demonstrate the power and varied utility of gene therapy approaches.

5.3.2 Arsenic

The metalloid arsenic is a naturally occurring toxicant found ubiquitously in the environment. In addition, it is a by-product of a number of industrial processes, including the production of copper and lead, as well as during burning of coal. Human exposure is thus by two different routes: ingestion through environmental exposure, usually via contaminated water sources, and occupational inhalation exposure via industrial processes. Chronic inhalation of arsenic results in an increased incidence of lung cancer, whereas chronic ingestion results in an increased incidence of skin, urinary bladder, liver and kidney cancer. The increased rate of cancer is associated with unequal segregation of chromosomes during cell division, meaning that arsenic may be classed as an aneugen. In addition, arsenic exposure has been observed to cause a wide range of cellular responses, including cell cycle arrest, cytoskeletal disruption, chromosomal aberrations and apoptosis. As seen with the aminoglycosides, toxicity within the kidney probably results from this organ's role in excretion. Concentration of arsenic prior to excretion occurs in the kidneys and urinary tract, and hence toxicity is situated there.

The exact mechanism underlying these changes is presently unknown, but is probably due to reactive oxygen and/or nitrogen species liberated during exposure to arsenic. To investigate the molecular mechanisms of arsenic toxicity Yih *et al.* (2002) used microarray technology to examine the changes in gene expression following exposure of human fibroblasts to arsenic. Using a microarray of 568 human genes Yih *et al.* identified 133 genes whose expression altered following acute exposure of human fibroblasts to arsenic. These genes could be clustered into six distinct groups dependent upon the temporal pattern and direction of change caused by exposure. Amongst the genes whose expression profiles were altered were several whose protein products are associated with transcriptional control, protein metabolism, cell cycle regulation and intracellular signalling. Identification of the changes in gene expression profiles following arsenic exposure provides important information towards the understanding of the molecular mechanisms underlying the established cellular responses, and has provided novel lines of investigation in understanding this toxicity.

5.4 Receptor-dependent toxicity

In the previous examples we have introduced the concept of chemical damage to cells occurring through their direct interaction with cellular components, or the action of their metabolites. However, most chemicals also interact with cellular molecules that transmit their effects through the cells, altering downstream processes; these are receptors.

As the name suggests a receptor may be defined as any molecule that a chemical can bind to and elicit a response from, thus segregating them from enzymes which alter the chemical itself upon binding, producing a metabolite. Due to this fairly open definition it is perhaps not surprising that receptors may carry out a great many different roles within the cell. As key regulators of cellular functions can toxic effects be mediated through them? To answer this question we will study two disparate receptor roles, and see how they may play a role in chemical toxicity.

5.4.1 Aryl hydrocarbon receptor (AhR)

The AhR was first identified in 1976 and has since been classified as a basic helix-loop-helix ligand-activated transcription factor. The *AhR* gene has been identified in all mammalian species studied and in some invertebrates, such as the nematode *C. elegans*. In common with other ligand-activated transcription factors it exists in the cytosol as a complex with a chaperone molecule, HSP90 in this case, when ligand is lacking. Binding of ligand causes a conformational change, disassociation from HSP90 and heterodimerization with the AhR nuclear translocator (ARNT). This complex then translocates to the nucleus, where it can bind to specific xenobiotic response elements (XRE) and activate gene expression (*Figure 5.6*). AhR is involved in the induction of a number of genes, which have become known as the aromatic hydrocarbon-responsive [Ah] gene battery (*Table 5.2*). These genes are involved in both Phase I and Phase II metabolism, and as we have already seen, induction of Phase I and Phase II enzymes expression often accompanies compound exposure; such a response helps to clear the compound quickly and efficiently from the body. Why then does induction via AhR often result in a toxic endpoint? There are probably two, related, answers to this question; firstly the ligands themselves and secondly the relative induction of the members of the [Ah] gene battery caused by them.

The ligands for AhR include many heterocyclic compounds that are present in the environment or are produced during industrial processes. These include the heterocyclic aromatic amines, formed during high-temperature cooking of meat and present in cigarette smoke, polycyclic aromatic hydrocarbons and polychlorinated biphenyls. The majority of these compounds form highly reactive intermediates during metabolism, and hence increased rates of metabolism will result in a higher frequency of adduct formation.

The second factor in the increased incidence of toxic effects associated with AhR activation is that increase in gene expression is not the same for all genes responsive to AhR. As seen previously, toxicity is often due not only to the production/presence of a reactive species, but also the rate of its removal from the body. It is this overall balance that determines whether

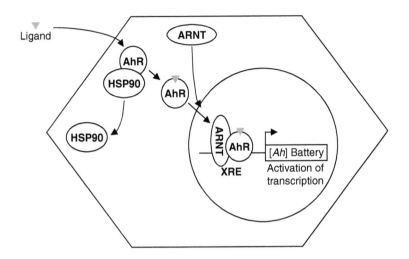

Figure 5.6

Transcriptional activation of the [Ah] gene battery.

toxic metabolites accumulate or not. In the case of the [Ah] gene battery expression of the Phase I CYPs is generally induced at lower concentrations of ligand than the other genes in the battery. For example, CYP1A1 gene expression is induced by 1000-fold lower concentrations of TCDD than UGT1A6 gene expression. Such differences would obviously lead to inequalities in the rate of production and removal of reactive intermediates and may increase the rate/extent of toxicity observed.

RayChaudhuri *et al.* (1990) examined the genes within the [Ah] gene battery and studied the inter-relationship between them, using engineered mutant strains of the hepatoma cell line Hepa1c1c7. They showed that removal of CYP1A1 activity, caused by a loss-of-function mutation in the CYP1A1 coding region resulted in increased levels of expression of the mutant CYP1A1 protein *and* of other genes within the battery. Addition of a functional CYP1A1 expression plasmid restored the level of expression to that seen in wild-type cells. This suggested that CYP1A1 is responsible for maintaining levels of an endogenous ligand of AhR, presumably through its metabolism. Without CYP1A1 activity, levels of this endogenous ligand increase, causing increased activation of AhR and therefore expression of genes in the AhR battery.

Table 5.2   The aryl hydrocarbon receptor (AhR) gene battery

Enzyme	Accession
CYP1A1	NM 000499.2
CYP1A2	NM 000761.2
NAD(P)H:quinine oxidoreductase (Nqo1)	NM 000903.1
Aldehyde reductase 3A1 (Aldh3A1)	NM 000691.3
Glucoronsyl transferase 1A6 (UGT1A6)	NM 001072.1
Glutathione A1 (GSTA1-1)	NM 145740.1

Members of the [Ah] gene battery are listed along with the RefSeq mRNA accession number for the human orthologue.

What, however, is the physiological role of AhR? The description so far classifies AhR as a responder to environmental toxins, and indeed one theory suggests that its origins were through the need to detoxify plant-derived toxins. However, evidence from evolutionary distant organisms suggests that AhR must perform endogenous functions as well. The *C. elegans* AhR is not activated by classical, mammalian ligands such as TCDD, yet its presence within the organism suggests it is required for a fundamental, biological role within *C. elegans*.

One method to understand the biological roles of proteins is through their targeted disruption in knock-out animals, and this approach has been tried for the AhR. AhR(−/−) transgenic mice were engineered, and study of them suggested endogenous roles for AhR in the development of the liver and vasculature, as these were both abnormally developed in the knock-out animals. However, no clear, precise, endogenous biological role has yet been ascribed to AhR.

An alternative method to knock-out studies for studying protein function is the over-expression of the protein, and Andersson *et al.* (2002) used this approach to attempt to delineate the physiological role of AhR. Mice engineered to contain a constitutively active AhR had a reduced life span, with lethality from 6 months of age, and were susceptible to stomach tumours. The location of the tumours is of interest as the breakdown products of several dietary constituents are indoles, and these have been suggested as putative endogenous ligands for AhR. The uncontrolled cell division associated with carcinogenesis provides further circumstantial evidence that AhR may play a role in regulating cell cycle control.

Study of the AhR gene battery may provide further clues to the endogenous role(s) of AhR. As the genes in the battery are involved in the metabolism of endogenous compounds as well as xenobiotics, co-ordinated regulation of the battery will affect this process. Whereas previously we have described the production of reactive intermediates and reactive oxygen species formed during uneven metabolism as a negative function, leading to toxicity, it is important to remember that reactive oxygen species are also cellular signalling molecules and are central to the regulation of cellular processes such as apoptosis and cell division. Nebert *et al.* (2000) have developed this hypothesis to produce a putative model of the roles of AhR battery genes in oxidative stress and cell signalling (*Figure 5.7*). AhR, in common with many ligand-activated transcription factors, may use alternate heterodimerization partners to regulate gene expression. AhR has been shown to interact with the retinoblastoma protein, and this complex acts as a suppressor of gene expression. The exact mechanism of this suppression is not known, but it does inhibit E2F-dependent transcription and cause cell cycle arrest (*Figure 5.7*; Puga *et al.*, 2000).

5.4.2 *N*-methyl-D-aspartate receptor (NMDA)

N-methyl-D-aspartate (NMDA) receptors are one of the body's major ligand-gated ionotropic receptors, regulating transport of Na^+, K^+ and Ca^{2+}. NMDA receptors are stimulated by glutamate, arguably the most important excitatory neurotransmitter in the brain. While such stimulation is vital to neuromodulation within the body, excess stimulation results in excitotoxicity and neuronal death, and NMDA-mediated excitotoxicity has been linked with a number of neurological disorders (*Table 5.3*).

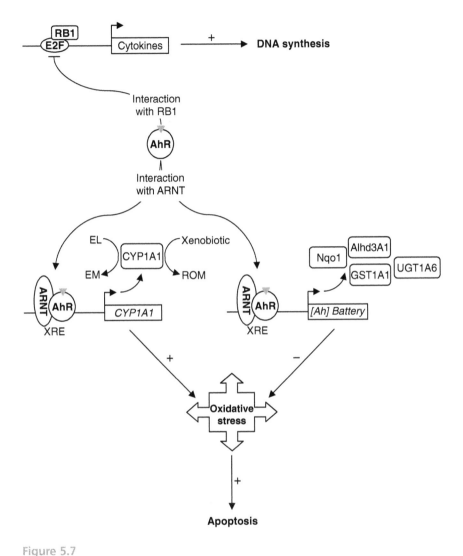

Figure 5.7

Role of AhR in cell signalling.

Functionally active NMDA receptors are composed of heterodimers, usually formed from a GRIN1 subunit (of which there is only a single gene but three alternate transcripts and hence protein products) and a GRIN2 subunit (of which there are four genes, *GRIN2A*, *GRIN2B*, *GRIN2C* and *GRIN2D*). Recently a third subunit family was identified, coded for by the *GRIN3A* and *GRIN3B* genes, although the function of these subunits is not clear at

Table 5.3 Neurological disorders linked to NMDA-mediated excitotoxicity

Acute seizure	Hyperglycinaemia
Acute ischaemia	Huntington's disease
Acute hypoglycaemia	HIV encephalopathy
Hyperammonaemia	Parkinson's disease

Table 5.4   NMDA-receptor subunits

Receptor subunit	Accession
GRIN1	NM 000832.4, NM 007327.1, NM 021569.1[a]
GRIN2A	NM 000833.2
GRIN2B	NM 000834.2
GRIN2C	NM 000835.2
GRIN2D	NM 000836.1
GRIN3A	NM 133445.1
GRIN3B	AC 004528.1[b]

GRIN, glutamate receptor, ionotropic, N-methyl D-aspartate. [a]GRIN1 has three alternative transcripts and RefSeq accession numbers are presented for each of these. [b]GRIN3B does not have a RefSeq accession number and the cDNA accession is presented instead.

present: NMDA subunits and their RefSeq mRNA accession numbers are presented in *Table 5.4*.

As stated above, toxicological issues exist with NMDA receptor agonists, as over-stimulation results in toxicity. NMDA-mediated excitotoxicity is complex for several reasons. Firstly, the stimulating agonist may vary; in general it is presumed to be glutamate but glycine or quinolinate may also play specific roles in activating this receptor. Secondly, glutamate itself may directly cause cell toxicity in some circumstances, and this must be distinguished from the NMDA-mediated toxicity elicited by glutamate. Finally, the result of NMDA receptor activation may be varied, ranging from the generation of reactive oxygen species (changes in gene expression and cellular damage; see Chapter 4 for details) to calpain activation (disruption of microtubule formation). An overview of the potential mechanisms of NMDA-mediated excitotoxicity is presented in *Figure 5.8* (Lynch and Guttmann, 2002).

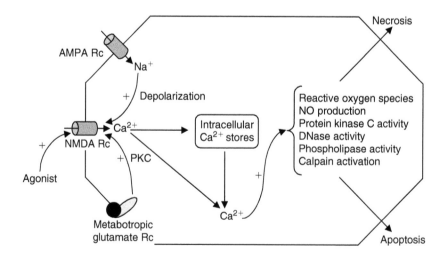

Figure 5.8

Potential mechanisms of NMDA-mediated excitotoxicity (adapted from Lynch and Guttmann, 2002).

A specific example of NMDA-mediated toxicity that has received considerable attention is ammonia toxicity. Ammonia is a naturally occurring
chemical which is formed within the body during the degradation of proteins. Under normal physiological conditions ammonia is removed from the
body before it can reach toxic levels through the action of the urea cycle,
incorporating ammonia into urea for excretion. However, in situations
where liver function is compromised, such as cirrhosis, blood ammonia
levels can rise to toxic levels and hepatic encephalopathy (liver-mediated
brain injury) results. Ammonia itself does not appear to be an agonist for
NMDA receptors, so how does it cause activation of NMDA receptors? One
possible explanation for these effects is that ammonia indirectly opens
NMDA ion channels. As can be seen from *Figure 5.8*, depolarization of cell
membranes may induce NMDA channel opening, by releasing Mg^{2+} from
the channel. Ammonia has been shown to depolarize membranes and it is
thus likely that it opens NMDA channels by this route (Monfort *et al.*, 2002).

5.5   Neurotoxicity

Compounds that cause neural toxicity may be categorized as those that
cause central nervous system (CNS) toxicity, those that cause peripheral
nervous system (PNS) toxicity and those that cause a mixture of the two.
While the body contains many enzymes to protect against potential toxins,
further steps are taken to protect arguably the most fragile organ in the
body, and the centre of the nervous system, the brain. As body functioning
is wholly dependent upon the brain then it is vital that no impairment of its
function occurs. Therefore, in addition to enzymes that react to potentially
harmful molecules, the brain uses a physical barrier to prevent toxin entry
into the brain: the blood–brain barrier.

Transport across most membranes in the body is aided by the presence of
multiple transport proteins and ion channels within the membranes, allowing routes for transport of molecules that are not sufficiently lipophilic to
cross membranes by passive diffusion. In addition there are collections of
'spacer proteins' between the individual cells forming 'gap junctions' through
which further transport can occur. In comparison, the cells of the blood–brain
barrier have a very restricted set of membrane proteins/ion channels and individual cells are connected by 'tight junctions' which do not allow transport.
The net result of this is that only a very restricted set of chemicals has ready
access to the brain, and this acts as the first line of defence to protect the brain
from potential toxins. However, as we will see below, a number of toxic chemicals are capable of crossing this barrier and causing damage to the brain.

5.5.1   Organophosphates

Organophosphates constitute probably the major class of insecticide compounds in use today. They are esters or thiols derived from phosphoric, phos-
phonic, phosphinic or phosphoramidic acids and are neurotoxic to both
mammals and insects, although they show greater toxicity towards the latter
group. In man, exposure is either occupational through skin absorption
during agricultural treatment, environmental through the contamination
of foodstuffs leading to ingestion or voluntarily through high level ingestion
of organophosphates during suicide attempts. Regardless of the route of
exposure, organophosphates exhibit two distinct modes of neurotoxicity,

an immediate inhibition of cholinesterase activity and a delayed neuropathy, both of which may be fatal and can occur after even a single exposure.

Organophosphates contain either a $P{=}S$ or $P{=}O$ moiety, with the latter being the active form; compounds containing $P{=}S$ moieties are converted to active $P{=}O$ containing metabolites via oxidative desulphuration during Phase I metabolism. Active $P{=}O$ containing organophosphates may interact with esterases in one of two ways, dependent upon the class of esterase that they interact with. Both class A and class B esterases catalyse the hydrolysis of organophosphates, however release of hydrolysed metabolites by class B esterases is extremely slow, and they are effectively irreversibly inhibited by organophosphates. Therefore, class A esterases hydrolyse organophosphates, resulting in their detoxification, whereas class B esterases are inhibited by organophosphates, and, as explained below, it is this that results in toxicity (*Figure 5.9*).

If inhibition of class B esterases causes the toxicity, study of these enzymes should clarify how this toxicity is brought about. Acetylcholinesterase is one of the predominant class B esterase enzymes and is responsible for the hydrolysis of the neurotransmitter acetylcholine. Such an action is important as it terminates the stimulation caused by acetylcholine and thus regulates the neurotransmission. Inhibition of acetylcholinesterase by organophosphates

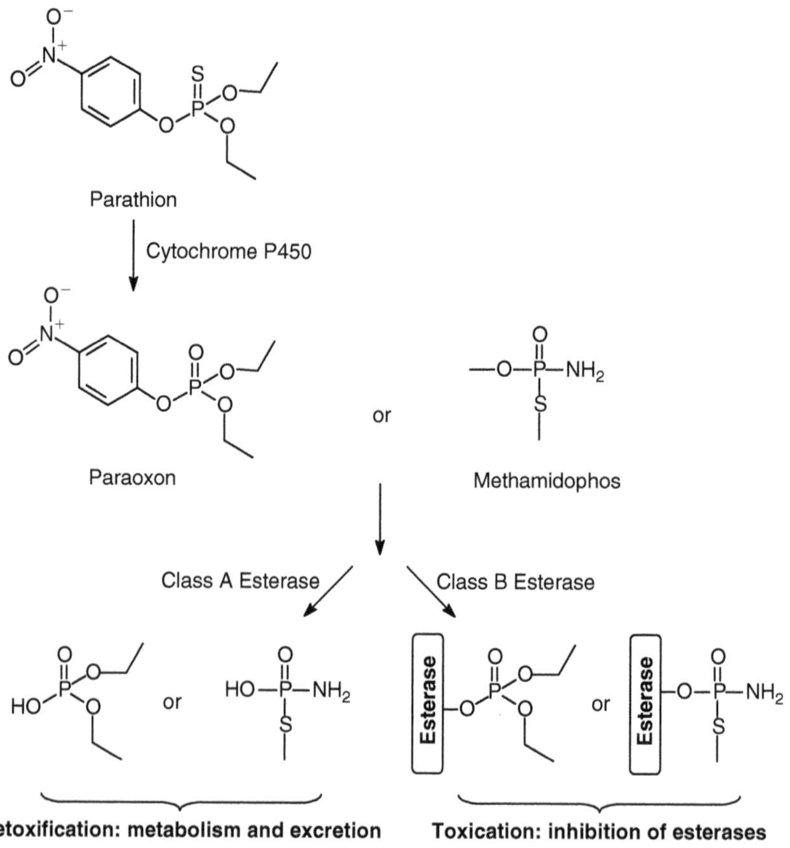

Figure 5.9

Toxicity of parathion.

results in a build-up of acetylcholine and the resultant over-stimulation of the nervous system causes excitotoxicity. Accumulation of acetylcholine affects both the peripheral and central nervous system in the human body, producing the classical symptoms of organophosphate poisoning. In the peripheral nervous system, over-stimulation of muscarinic receptors in the smooth muscle, heart and exocrine glands leads to bronchoconstriction and bradycardia. In addition, action on nicotinic receptors in skeletal muscle results in involuntary twitching and general muscular weakness. Build-up of acetylcholine within the central nervous system leads to a variety of symptoms including tension, anxiety, convulsions and reduced levels of respiration and cardiac output. Once inhibition of cholinesterases within the human body approaches 80% death is imminent, usually through respiratory failure.

The second toxicity associated with organophosphates is delayed neuropathy. Whereas cholinesterase inhibition-mediated effects present within 48 hours of exposure, delayed neuropathy does not become evident until 10–14 days following exposure. The mechanism of this neuropathy is quite distinct from the acute, cholinesterase-mediated, effects and involves the phosphorylation of another class B esterase within the nervous system, termed the neuropathy-targeted esterase (NTE). NTE appears to play a role in hydrolysis of membrane lipids, and phosphorylation of NTE results in its inhibition, which in turn leads to the symptoms of neuropathy, including paralysis and ataxia (Lotti, 1992).

We have seen in this and previous chapters the use of transgenic mouse models to examine the role of target proteins in response to a toxic chemical. Through the use of knock-out or knock-in transgenic animals it is possible to study not only the role of individual proteins, but also inter- and intra-species differences. There is, however, an alternate use of transgenic animals, their application to the determination of the site and extent of genotoxicity caused by toxins. Muta™Mouse is a transgenic mouse model in which the bacterial *lacZ* gene is integrated into the mouse genome. Exposure to mutagens causes damage to DNA, and results in disruption of the *lacZ* gene. Later detection in an *in vitro* system allows a quantitative determination of the mutation frequency in any tissue to be made, thus allowing the sites of toxicity caused by a chemical to be assessed. This technology was used to examine if organophosphate toxicity caused DNA mutations in addition to the esterase inhibition previously described. Plesta *et al.* (1999) exposed λlacZ transgenic mice to the organophosphate Dichlorvos for acute (4 hours) or chronic (14 days) exposures and then looked for the induction of somatic mutations. Following acute exposure, no evidence of increased somatic mutations was seen, but following chronic exposure a three-fold increase in this rate was observed. Such information supported previous *in vivo* studies that suggested that mutagenesis by Dichlorvos is possible under prolonged exposure. However, Dichlorvos is not thought of as a carcinogen; why? As we have already seen, organophosphate toxicity caused by esterase inhibition can manifest after a single dose. Hence, while there is a hazard of mutagenesis from Dichlorvos following chronic exposure, any effects are masked by the esterase-inhibition-mediated neurotoxicity caused by a single, acute exposure.

An interesting recent development is the possibility of gene therapy to circumvent organophosphate poisoning (Cowan *et al.*, 2001). The body has a natural defence mechanism against organophosphates, hepatic and serum paraoxonase, an arylesterase enzyme that can hydrolyse organophosphates, resulting in their detoxification. Cowan *et al.* hypothesized that artificial increases in the level of paraoxonase would result in the hydrolysis of organophosphates in the blood and liver, detoxifying them before they could enter the nervous system and cause neurotoxicity. Using an adenovirus delivery system, they expressed increased levels of human paraoxonase in the serum of mice and then examined their response to the organophosphate chlorpyrifos. Using this gene delivery system they were able to boost serum arylesterase levels by up to 60% and, more importantly, significantly decreased the chlorpyrifos-mediated inactivation of brain acetylcholinase. This resulted in a general protection of the mice against the toxic effects of organophosphates and suggested that paraoxonase could be a novel therapeutic target for the treatment of organophosphate poisoning. Drugs targeted to increase levels of endogenous paraoxonase activity would have the same effect as the retroviral system and increase serum arylesterase levels and hence increase the rate of organophosphate detoxification.

5.5.2 MPTP

Parkinson's disease is one of the most common neurodegenerative disorders to occur in man. It was first categorized in 1817 by Dr J. Parkinson, from whom it derives its name. Research into parkinsonism was slowed by the lack of an appropriate animal model; without such models it is difficult to study the molecular mechanisms behind any disorder as well as to identify and test novel therapeutic compounds. However, in 1985 several individuals were reported to show Parkinson-like symptoms following the use of a synthetic heroin substitute. Analysis of the synthetic heroin revealed the presence of 1-methyl-4-phenyl-1,2,3,6-tetrahydropyridine (MPTP), a by-product of 1-methyl-4-phenyl-4-propionoxypiperidine (MPPP) synthesis (*Figure 5.10*;

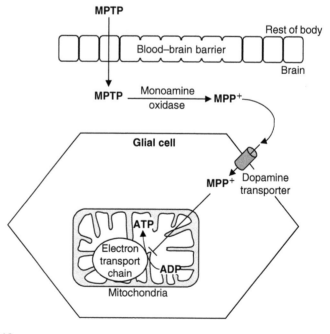

Figure 5.10

Toxicity of MPTP.

Ballard *et al.*, 1985). The pathological changes caused by MPTP, atrophy and loss of pigment-containing neurons in the substantia nigra and locus ceruleus of the brain, as well as the clinical presentation, matched those of parkinsonism. With this chance discovery the scientific community gained an important tool in the delineation of the molecular mechanisms of parkinsonism as mice treated with MPTP provided the first animal model of the disease.

MPTP readily crosses the blood–brain barrier and enters into the brain. Once there, it is a substrate for monoamine oxidase, which converts it into the ultimate toxic metabolite 1-methyl-4-phenylpyridium (MPP^+). MPP^+ is then readily taken up by glial cells (specialized cell types which surround the neurons in the brain) via the dopamine transporter, and it then accumulates in the mitochondrial matrix of these cells. Accumulation of MPP^+ results in inhibition of Complex I of the electron transport chain (NADH CoQ1 reductase) and results in cell death due to ATP depletion (*Figure 5.10*).

Sriram *et al.* (2002) demonstrated that levels of TNFα, a pro-inflammatory cytokine, increased following exposure of mice to MPTP and thus examined the role of this signal transduction protein in MPTP toxicity. Transgenic mice were engineered to lack the receptor for TNFα, and hence effectively block signal transduction caused by TNFα. TNFα acts on two receptors in the brain and therefore both single TNF receptor knockouts (TNFR1 $-/-$ or TNFR2 $-/-$) and double receptor knockouts (TNFR-DKO) were engineered and examined for how the loss of these receptors affected MPTP-mediated toxicity. Ablation of one receptor provided some protection from the effects of MPTP, reducing changes in three markers of MPTP-mediated toxicity (striatal dopamine, tyrosine hydroxylase and glial fibrillary acidic protein levels) to half that seen in the wild-type ($+/+$) mice. Even better, TNFR-DKO mice were completely protected from the effects of MPTP, implicating the signal transduction effects of TNFα as an obligatory for MPTP-mediated neurotoxicity. This presents the possibility of novel therapeutics for the treatment of parkinsonism, based upon modulation of the TNFα signalling pathway.

The discovery of MPTP as a model of Parkinson's disease has also focused efforts to deduce the natural causes of parkinsonism. While an increasing number of papers suggest a genetic component, resulting in familial parkinsonism, the occurrence of non-familial parkinsonism and the fact that familial parkinsonism does not show complete penetrance suggest an external trigger may exist. Based upon the evidence of MPTP there are two potential sources for such a trigger; structures chemically related to MPTP, or structurally unrelated compounds that have similar biological effects.

1,2,3,4-tetrahydroisoquinoline (TIQ) derivates are present in many foodstuffs (e.g. milk, banana) and hence present a possible environmental source of MPTP-related compounds. TIQ may be metabolized in the brain by 2-methyltransferase, and the resulting compound (2-methyl TIQ) produces similar effects on dopaminergic glial cells as MPTP (*Figure 5.11*; Fukuda, 2001). An alternative, environmental, trigger for parkinsonism may exist in compounds that are chemically unrelated to MPTP but which are also capable of disrupting complex I of the electron transport chain. Rotenone, an agricultural insecticide, has long been known to be a high-affinity, specific inhibitor of complex I of the electron transport chain and therefore is a candidate for an environmental trigger of Parkinson's. Indeed, exposure of mice to rotenone does result in pathologically similar effects to those seen in Parkinson's, with a decrease in pigmented neurons in the substantia nigra, and accumulation of α-synuclein (central to the formation of Lewy bodies, a classical sign of parkinsonism).

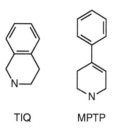

TIQ   MPTP

Figure 5.11

TIQ derivatives: environmental counterparts of MPTP?

Thus, through the study of a unique neural toxin, MPTP, large steps have been made in the understanding of Parkinson's disease, both in terms of its underlying mechanisms, the most effective treatment and possible environmental cues for its appearance.

5.6 Teratogenesis

Teratogenesis, the abnormal development of a fetus, is a complex subject, dependent upon the myriad of changes that occur during this period in development. The risk of a teratogenic event occurring during a pregnancy has been estimated at approximately 3–5%, although this can be elevated in any individual through either genetic predisposition or exposure to teratogenic compounds. Compounds may be directly teratogenic, or require maternal activation before becoming toxic. Toxicity may also be dependent upon the route of exposure, as this may alter the amount of compound distributed to the fetus, and on the developmental stage of the fetus when it is exposed to the toxicant.

In general the fetus is most susceptible to teratogens during organogenesis, the period 18–60 days post-conception when the major organs are formed. Teratogens can cause abnormalities in cell division, migration or apoptosis, affecting overall histogenesis or altering gene expression within individual groups of cells, changing cellular fates. *Table 5.5* lists some clinically used compounds recognized to possess teratogenic potential.

5.6.1 Anticonvulsant drugs

One simple method of preventing xenobiotic-mediated teratogenesis is to limit the exposure of pregnant mothers to compounds that are known to be teratogens. While this is relatively simple in most cases, treatment of long-term illness presents a special complication. In such cases, treatment may be required to continue throughout the period of pregnancy, and thus the risks to the developing fetus must be assessed. One such situation is in the treatment of epilepsy, and it has been estimated that the risk of teratogenesis can be as much as ten times higher for mothers taking antiepileptic drugs compared to individuals not receiving these drugs. Risk assessment is further complicated by the frequent use of polypharmacy in the treatment of epilepsy. Lindhout *et al.* (1984) examined the relative risks of teratogenesis in epileptic mothers taking between one and four anticonvulsant drugs. They showed that a patient taking no anticonvulsant drugs had a 2% risk of

Table 5.5 Drugs with teratogenic potential

Drug (class)	Therapeutic target	Teratogenic effects
ACE inhibitors	Hypertension	Oligohydramnios; intrauterine growth retardation; renal failure; hypotension; pulmonary hypoplasia; abortion
Antiepileptic	Anticonvulsant	CNS, cardiac, eye, GIT and genitourinary defects; facial dysmorphism; digital hypoplasia; growth retardation
Cyclophosphamide	Cancer	Skeletal and ocular defects; cleft palate
Danazol	Endometriosis	Masculinization of female external genitalia
Diethylstilbestrol	'Morning after' pill	Genital carcinoma (females); genital tract abnormalities (male and female)
Lithium	Mental illness	Cardiac defects
Quinine	Malaria	Deafness; abortion
Thalidomide	Immunopathologic diseases	Limb reduction; cardiac, urogenital, renal, orofacial, ocular and GIT defects; cranial nerve anomalies, microtia

Adapted from Polifka and Friedman (2002).

giving birth to offspring with congenital abnormalities, while treatment with a single anticonvulsant compound only increased that risk to 3%. However, polypharmacy increased the risk to 5, 10 and 20% for women taking two, three or four drugs in combination. Such data suggest that during pregnancy one potential course of action is the limitation of polypharmacy, with its vastly increased risks of teratogenesis. By using single-drug therapy the risks to both mother and developing fetus are thus minimized.

5.6.2 Thalidomide

Thalidomide is perhaps one of the most notorious drugs developed in the past 60 years. Thalidomide was first introduced in the late 1950s in Europe and Canada as a sedative and was widely prescribed to treat nausea and insomnia in pregnant women. However, it soon became apparent that its use in the first trimester of pregnancy led to severe birth defects, and this resulted in its withdrawal from the market in 1961.

Thalidomide teratogenesis is manifested as severe limb reduction, often accompanied by other organ defects (*Table 5.5*). Despite the large number of scientific papers published on this subject over the past 60 years a definitive mechanism for thalidomide teratogenesis has not been produced. Of the many hypotheses presented it is probable that the underlying mechanism affects one of the following processes: synthesis/function of growth factors or integrins, angiogenesis, chondrogenesis, DNA synthesis or apoptosis (Stephens and Fillmore, 2000). Indeed, it seems probable that the mode of action is a combination of all of the above effects, and this has led Stephens *et al.* (2000) to propose a generalized mechanism for its teratogenic effects (*Figure 5.12*). Thalidomide is thought to bind to GC-rich DNA sequences and, in the model of Stephens *et al.*, thalidomide therefore competes with the transcription factor SP1 for its binding site (GGGCGG). Such binding would reduce SP1-mediated gene expression, potentially causing the observed effects. However, SP1 is a ubiquitous transcription factor

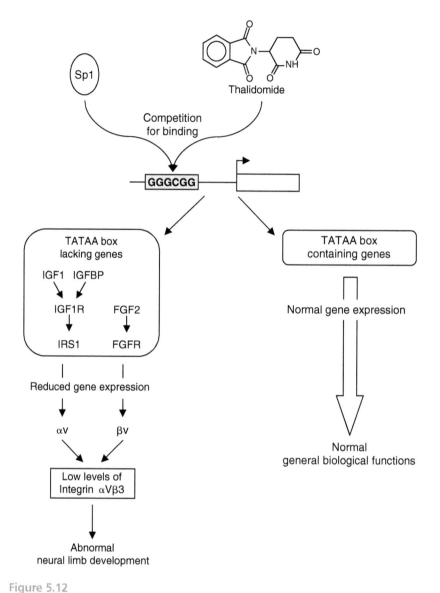

Figure 5.12

Proposed mode of thalidomide toxicity (adapted from Stephens *et al.*, 2000).

involved in the regulation of a large number of genes; how then is the tissue specificity of thalidomide toxicity explained in this model? As discussed in Chapter 7, over 90% of mammalian genes contain TATAA boxes within their promoters, and these sites are involved in the efficient recruitment of the RNA polymerase complex to the transcription start site. Stephens *et al.* proposed that in these genes competition at the SP1 site may have little visible effect, and transcription would continue as normal. However, in genes *without* TATAA boxes, SP1 binding plays a critical role in recruitment of the transcription complex and hence effects would be more pronounced in these genes. The overall effect of this would be that genes *with* TATAA boxes would have normal levels of gene expression, whereas genes *without* TATAA boxes would have significantly lower levels of expression. Amongst the genes known to lack TATAA boxes are those coding for factors involved in the tissue-specific expression of integrins (molecules important in cell motility). As integrins play a fundamental role in angiogenesis, Stephens *et al.* hypothesized that thalidomide-mediated inhibition of gene expression would result in reduced integrin production. This, in turn, would cause the abnormal limb development associated with thalidomide toxicity.

The mechanism of thalidomide-mediated teratogenesis has become an important issue in the past few years with the re-emergence of thalidomide as a therapeutic tool. In 1997, the United States Food and Drug Administration approved thalidomide for treatment of erythema nodosum leprosum, an acute, debilitating, phase of leprosy. Since that time, the use of thalidomide for several other disorders, including weight loss in tuberculosis, ulcers and weight loss associated with HIV and cancer has been proposed. Despite the extremely tight regulations placed upon thalidomide usage by the regulatory bodies, new cases of teratogenesis are emerging as the use, and occasional misuse, of thalidomide increases. If the exact nature of the teratogenic mechanism could be delineated it might be possible to ascertain which portion of the thalidomide chemical structure is responsible for these effects. A related drug could then be designed lacking these chemical signatures and resulting in compounds still maintaining efficacy but with fewer adverse effects, therefore increasing the utility of thalidomide-type drugs in these exciting new therapeutic areas.

5.7 Summary

In this chapter we have examined some of the varied modes of action of chemicals in causing toxicity. We have seen that even relatively simple chemical structures, such as carbon tetrachloride, can precipitate a wide range of effects within the body, leading to gross toxicity if left unchecked. We have also seen that modern molecular toxicology is beginning to unravel the molecular mechanisms underlying toxic events, and thus increase our understanding of the overall biological response to chemicals, both in a 'chemical-specific' manner and a more generalized 'chemical class-specific' manner. Such information will be vital for the development of therapies to treat and/or prevent such toxic effects, as well as the design of novel therapeutic drugs with all the good, therapeutic, effects of their predecessors, but without the associated side effects.

So far we have considered the effects of exposure to a single compound only. In the next chapter we will examine the effects of exposure to mixtures of compounds, and consider how we form realistic risk assessments of such mixtures.

References

Andersson, P., McGuire, J., Rubio, C., *et al.* (2002) A constitutively active dioxin/aryl hydrocarbon receptor induces stomach tumours. *Proc. Natl. Acad. Sci. USA* **99**(15): 9990–9995.

Arni, P. and Hertner, T. (1997) Chromosomal aberrations in vitro induced by aneugens. *Mutation Res.* **379**(1): 83–93.

Aubrecht, J., Goad, M.E.P., Simpson, E.M., *et al.* (1997) Expression of *hyg^R* in transgenic mice causes resistance to toxic effects of hygromycin B in vivo. *J. Pharmacol. Exp. Therapeutics* **281**: 992–997.

Ballard, P.A., Tetrud, J.W. and Langston, J.W. (1985) Permanent human parkinsonism due to 1-methyl-4-phenyl-1,2,3,6-tetrahydropyridine (MPTP): seven cases. *Neurology* **35**(7): 949–956.

Cowan, J., Sinton, C.M., Varley, A.W., *et al.* (2001) Gene therapy to prevent organophosphate intoxication. *Toxicol. Appl. Pharmacol.* **173**(1): 1–6.

Fukuda, T. (2001) Neurotoxicity of MPTP. *Neuropathology* **21**: 323–332.

Hodgson, E., Mailman, R. and Chambers, E. (1998) *Dictionary of Toxicology*, 2nd edn. Macmillan Reference Ltd., London.

Joseph, P., Lei, Y.X., Whong, W.Z., *et al.* (2002) Oncogenic potential of mouse translation elongation factor-1 delta, a novel cadmium-responsive proto-oncogene. *J. Biol. Chem.* **277**(8): 6131–6136.

La, D.K. and Swenberg, J.A. (1996) DNA adducts: biological markers of exposure and potential applications to risk assessment. *Mutation Res.* **365**: 129–146.

Lazo, J.S., Kondo, Y., Dellapiazzi, D., *et al.* (1995) Enhanced sensitivity to oxidative stress in cultured embryonic cells from transgenic mice deficient in metallothienein I and II genes. *J. Biol. Chem.* **270**: 5506–5510.

Lindhout, D., Hoppener, R.J. and Meinardi, H. (1984) Teratogenicity of antiepileptic drug combinations with special emphasis on epoxidation (of carbamazepine). *Epilepsia* **25**(1): 77–83.

Lotti, M. (1992) The pathogenesis of organophosphates. *Crit. Rev. Toxicol.* **21**: 465–487.

Lynch, D.R. and Guttmann, R.P. (2002) Excitotoxicity: perspectives base on N-Methyl-D-Aspartate receptor subtypes. *J. Pharmacol. Exp. Therapeutics* **300**: 717–723.

Monfort, P., Kosenko, E., Ereg, S., *et al.* (2002) Molecular mechanism of acute ammonia toxicity: role of NMDA receptors. *Neurochem. Int.* **41**: 95–102.

Nebert, D.W., Roe, A.L., Dieter, M.Z., *et al.* (2000) Role of the aromatic hydrocarbon receptor and [*Ah*] gene battery in the oxidative stress response, cell cycle control, and apoptosis. *Biochem. Pharmacol.* **59**: 65–85.

Pletsa, V., Steenwinkel, M.J., van Delft, J.H., *et al.* (1999) Induction of somatic mutations but not methylated DNA adducts in lambdalacZ transgenic mice by dichlorvos. *Cancer Lett.* **146**(2): 155–160.

Polifka, J.E. and Friedman, J.M. (2002) Medical genetics: 1. Clinical teratology in the age of genomics. *J. Can. Med. Assoc.* **167**(3): 265–273.

Puga, A., Barnes, S.J., Dalton, T.P. *et al.* (2000) Aromatic hydrocarbon receptor interaction with the retinoblastoma protein potentiates repression of E2F-dependent transcription and cell cycle arrest. *J. Biol. Chem.* **275**(4): 2943–2950.

RayChaudhuri, B., Nebert, D.W. and Puga, A. (1990) The murine *Cyp1a1* gene negatively regulates its own transcription and that of other members of the aromatic hydrocarbon-responsive [*Ah*] gene battery. *Mol. Endocrinol.* **4**: 1773–1781.

Simeonova, P.P., Gallucci, R.M., Hulderman, T., *et al.* (2001) The role of tumor necrosis factor-alpha in liver toxicity, inflammation, and fibrosis induced by carbon tetrachloride. *Toxicol. Appl. Pharmacol.* **177**(2): 112–120.

Sriram, K., Matheson, J.M., Benkovic, S.A., *et al.* (2002) Mice deficient in TNF receptors are protected against dopaminergic neurotoxicity: implications for Parkinson's disease. *Faseb J.* **16**(11): 1474–1486.

Stephens, T.D. and Fillmore, B.J. (2000) Hypothesis: thalidomide embryopathy – proposed mechanisms of action. *Teratology* **61**: 189–195.

Stephens, T.D., Bunde, C.J.W. and Fillmore, B.J. (2000) Mechanism of action in thalidomide teratogenesis. *Biochem. Pharmacol.* **59**: 1489–1499.

Swenberg, J.A., Ham, A., Koc, H., *et al.* (2000) DNA adducts: effects of low exposure to ethylene oxide, vinyl chloride and butadiene. *Mutation Res.* **464**(1): 77–86.

Wang, X., Lau, M., Shi, Y.Q., *et al.* (2002) Differential display of vincristine-resistance-related genes in gastric cancer SGC7901 cell. *World J. Gastroenterol.* **8**(1): 54–59.

Waring, J.F., Ciurlionis, R., Jolly, R.A., *et al.* (2001) Clustering of hepatotoxins based on mechanism of toxicity using gene expression profiles. *Toxicol. Appl. Pharmacol.* **175**(1): 28–42.

Yih, L.H., Peck, K. and Lee, T.C. (2002) Changes in gene expression profiles of human fibroblasts in response to sodium arsenite treatment. *Carcinogenesis* **23**(5): 867–876.

The real world – complex mixtures

6

6.1 Introduction to complex mixtures

Much of the discussion within this book has so far considered only the simplest case of toxicity, exposure to a single chemical. Indeed, much of the information used to investigate mechanisms of toxicity and subsequent risk assessment is based upon such a paradigm. However, in the real world this simple situation is seldom seen. The action of any one chemical on the body, whether that is beneficial or deleterious, may be altered by the presence of other chemicals in the body at the same time. In the most studied cases one, or both, of these chemicals is a clinical drug and interactions may lead to loss of therapeutic efficacy or toxicity (an adverse drug response or ADR).

Exposure to mixtures is not only limited to therapeutic medicines, and mixtures of environmental pollutants, food additives and the like form a substantial part of mixtures that humans are exposed to every day. Such mixtures may, however, be harder to predict as the combinations we are exposed to are not as fixed, depending instead upon our lifestyles; a non-smoking, teetotal vegan living in the middle of the Sahara will be exposed to very different chemicals than a smoking, alcohol-drinking carnivore living in the middle of London. This chapter will discuss the need to address such issues and expand on the strategies developed to carry this out.

Is the exposure to multiple chemicals at any one time significantly different from the total effect of the individual chemicals? On the whole the answer is probably no, with the individual chemicals impacting upon the effects of other chemicals in negligible ways, or not at all. In this situation, the overall effect can be calculated as the sum of the individual effects, and indeed this additive approach is the default paradigm for risk assessment. However, a significant number of chemical combinations do result in reactions significantly different from a simple additive effect, either causing an increase in the overall effect (synergy) or decrease (antagonism); it is these that cause concern in terms of risk assessment. A recent study by McDonnell and Jacobs (2002) studied the rate of hospital admissions in the USA due to adverse drug reactions, and subdivided them into the causes of the ADR. While the largest cause for adverse reactions was inadequate monitoring of drug therapy (67%), 26% of the adverse reactions could be directly attributed to drug–drug interactions. If one considers that up to 28% of hospital admissions in the USA are due to adverse drug reactions then it becomes clear that such interactions play an important role in patient healthcare in the Western world. There therefore exists an obvious need for experimental strategies to assess the effects of exposure to chemical mixtures, thus allowing accurate risk assessments to be made.

Chemical mixtures can be subdivided into two categories, simple and complex, and these have slightly different methods of assessment. A simple mixture may be defined as any mixture composed of ten or less chemicals, the composition of which is known both qualitatively (i.e. what chemicals are in the mixture) and quantitatively (i.e. how much of each chemical is present in the mixture). Examples of simple mixtures include poly-pharmacy of drug cocktails or the interaction of therapeutic drugs with non-medicinal drugs, such as herbal remedies. In comparison, complex mixtures are composed of tens, or even thousands, of chemicals and the exact composition of the mixture may be unknown in either a qualitative or quantitative fashion; examples of such mixtures include environmental air pollution, drinking water treatments or natural flavouring complexes.

6.2 Simple mixtures

6.2.1 Analysis methodologies

Simple mixtures, with ten or less components, are, both conceptually and practically, the most straightforward type of mixture to analyse. The exact contents of the mixture are known, both in terms of which chemicals are present but also in what proportions. To assess whether exposure to this mixture produces effects significantly different from purely additive, two approaches can be taken. Firstly, the entire mixture can be treated as a single entity, and be examined using a standard battery of toxicity tests, so that the hazard, and ultimately therefore risk, can be characterized for the entire mixture. This approach has the advantage that whole-mixture effects are assessed but the distinct disadvantage that it does not identify which of the components of the mixture are responsible for any effects or interactions. Because of this, any new mixture will have to be assessed from scratch, even if it differs from the original mixture by only a single component as no information will be available on the nature of any interactions of chemicals within the mixture which could be used to extrapolate and identify potential interactions in this second mixture.

A better approach therefore is to test the components, both individually and in combination with each other. Such an undertaking is not a minor one however: to test all the interactions in a five-compound mixture equates to five single compound tests plus 120 combinations. Doubling this to ten compounds does not double the number of tests required but instead raises it to over 3.5 million test sets, well beyond the scope of most facilities. How then do you reduce this to a realistic number of combinations to study? One way is through the use of historical data to allow a mathematical analysis of the mixture and thus predict the overall hazard of the mixture. The default analysis carried out assumes no interactions occur within the mixture, and thus the overall effect is the sum of the individual effects (i.e. additive). Hazard Index (HI), as proposed by the US Environmental Protection Agency, takes such an approach (EPA, 1986). In this system, each individual chemical has a HI value, based upon historical data and the overall hazard of the mixture is calculated by adding these, and from the cumulative HI a risk assessment can be made. However, such an analysis does not take into account any interactions within the mixture and hence does not meet the full requirement; i.e. the ability to test if a chemical mixture results in significantly different effects on the body than if the components were given separately. Mumtaz and Durkin (1992) described a refinement of the HI system, incorporating weight of evidence (WoE) for chemical interactions. Here, a modifier for the HI is derived, based upon known interactions between chemicals within the mixture. This modifier is dependent upon the published literature and includes factors such as route of exposure and modifiers of the interaction, such as other chemicals within the mixture. Does such a weighting system work? To test this Mumtaz et al. (1998) examined the calculated toxicity of two simple mixtures using the WoE approach and compared this to experimentally derived data. The first group consisted of four halogenated aliphatics (trichloroethylene, tetrachloroethylene, hexachloro-1,3-butadiene and 1,1,2-trichloro-3,3,3-trifluoropropene) which are known to cause nephrotoxicity via the same general mechanisms. With this group, the WoE approach closely predicted the cumulative effect of the four compounds when given as a simple mixture. The second group consisted of four nephrotoxins with differing mechanisms of action (mercuric oxide, lysinolalanine, D-limonene and hexachloro-1,3-butadiene), and with this simple mixture the WoE approach did not model the experimental data well. This is probably due to the presumptions used in WoE calculations, which state that interactions tend to occur between chemicals with shared modes of action. Hence, the WoE system provides an improvement on the simple HI system, but with the caveat that it works best with mixtures of similar chemicals.

An alternative to the WoE modified HI approach of Mumtaz and Durkin is the toxic equivalency factor (TEF; Van den Berg, 1998). As we saw with the WoE system, estimation of overall effect of a mixture works best for mixtures where the mechanisms of toxicity are similar for the chemicals within the mixture. Using this presumption, structurally related chemicals are assigned a TEF based upon their toxicological impact relative to the most potent chemical in their group. The overall effect of a mixture of these chemicals is then calculated as the sum of 'concentration $\times$ TEF' for each chemical. An example of a chemical class where TEFs have been successfully used is in the classification of polychlorinated dibenzo-p-dioxins (PCDDs), which are environmental contaminants and may be potent carcinogens. Table 6.1 shows the TEFs derived for several PCDDs, relevant to the most toxic chemical 2,3,7,8 TCDD (TEF = 1).

TEFs use the same basic presumptions as the WoE approach, in that they presume similar chemicals have approximately the same pharmacokinetics, and act through the same receptors/activation pathway. It can be seen therefore that, as with the WoE approach, TEFs are appropriate for certain mixtures

Table 6.1 Toxic equivalency factors assigned to dioxins for mammalian risk assessment

Chemical	TEF
2,3,7,8-TCDD	1
1,2,3,7,8-PentaCDD	1
1,2,3,4,7,8-HexaCDD	0.1
1,2,3,6,7,8-HexaCDD	0.1
1,2,3,7,8,9-HexaCDD	0.1
1,2,3,4,6,7,8-HeptaCDD	0.01
OctaCDD	0.0001

Adapted from Van den Berg et al. (1998).

such as environmental pollutants which are formed of many structurally related chemicals but are not appropriate for more diverse mixtures. TEFs have been increasingly used in risk assessment, and they seem to provide a valid system for calculating the overall effect of a mixture. However, as with all evidence-based systems, further data must be used to constantly re-validate the presumptions of the system, and to test the TEF concept.

Regardless of which mathematical system is used for calculating the overall effect of a simple mixture, we need to generate the physical data that form both the literature base for such mathematical formulae, and is imperative to validate them. Such data can be either prospective, for interactions that we predict will occur, or retrospective, for non-predictable interactions. In the former case, an individual is known to be exposed to both chemicals and hence the possibility of an interaction can be predicted. Probably the most obvious case of such circumstances is during poly-pharmacy, where an individual may be given several drugs at one time and hence interactions of the individual drugs must be calculated to ensure full therapeutic benefit for all of them, without producing an ADR.

In comparison, unpredictable interactions occur when one or more of the chemicals is not consciously given to an individual. Examples of such interactions often exist when individuals are on a single drug therapy and are then exposed to a second chemical that causes an interaction. The possible effects of such interactions must then be assessed to calculate potential risk to the individual, both in terms of direct toxicity and lack of therapeutic effect of the drug.

6.2.2 Assessment of interactions

An example of an experimental system designed to examine interactions within a simple mixture was presented by Tajima *et al.* (2002), who used an *in vitro* assay to test for interactions between a group of mycotoxins. Mycotoxins are an important contaminant of foodstuffs, in particular those produced by the fungal genus *Fusarium*. As a group, the *Fusarium*-derived mycotoxins are structurally and toxicologically diverse, and an initial assessment might conclude that they are unlikely to interact, as one predictor of chemicals that may interact is shared features, be they structural, metabolic or toxicological. However, as infested foodstuffs are always contaminated as a mixture it is important to assess if such an initial hypothesis is valid, and therefore risk assessments can be made solely on the basis of the effects of the individual components, or if interactions do exist and must be taken into account during risk assessment.

Due to the technical complexity of studying multiple chemical combinations, such analysis lends itself to high-throughput *in vitro* assays, which allow the rapid assessment of potential interactions – these can then be studied in further detail. Tajima *et al.* used a high-throughput assay, in which mouse fibroblast L-929 cells were seeded into 96-well plates and exposed to five *Fusarium* mycotoxins either alone or in combination; toxicity was assessed by inhibition of DNA synthesis within the cells. Due to the rapidity of the screen it was possible to screen many combinations of the mycotoxins, altering their relative ratios, and thus examine for any interactions that may be occurring.

Four of the five mycotoxins were able to significantly inhibit DNA synthesis and, more importantly, combinations of the mycotoxins inhibited DNA synthesis more severely than would be suggested by a purely additive effect (i.e. they showed a synergy of action). Further work could then be carried out on the mycotoxin combinations to determine the exact nature of these interactions. Thus, it was possible to not only detect interactions in a rapid-throughput *in vitro* system, but also to identify which components were responsible for the interactions. Such experimental approaches may represent a significant step forward in assessing simple mixtures and assuring a more accurate risk assessment for human health.

6.2.3 Predicted mixtures

Perhaps the most obvious mixtures are those deliberately given to an individual; they are thus predictable. As we know that the individual is going to be exposed to this particular combination then it is appropriate that risk assessment is made based upon the mixture and not the individual components. Combination chemotherapy is used increasingly in medial healthcare today, and this is especially true for complex, life-threatening diseases such as HIV or cancer. In HIV treatment, combinations of antiretroviral drugs, usually either protease inhibitor or nucleoside/non-nucleoside reverse transcriptase inhibitors are often used. As each drug will have different strengths and weaknesses, for example having differing penetrations into differing compartments within the body, such combination therapies have provided new levels of success in the treatment of HIV. However, the addition of combination therapies to an already complex treatment regime dramatically increases the chances of drug–drug interactions, and such interactions must be assessed to ensure no loss of efficacy, or increase in toxicity is observed.

An example of the potential implications of such combination therapy was highlighted by Veldkamp *et al.* (2001), who studied the effect of two non-nucleoside reverse transcriptase inhibitors (NNTIs) when used in combination. As described above, a combination usually only becomes likely to cause an interaction if the two chemicals show similar modes of action and/or metabolism; both nevirapine and efavirenz, the NNTIs under study, have the same overall mechanism of action, preventing the functioning of reverse transcriptase. In addition, both chemicals undergo CYP-mediated Phase I metabolism, with specificity towards CYP3A4 and CYP2B6, and hence may interfere with the metabolism of each other. Hence, it may be predicted that use of nevirapine and efavirenz in combination will lead to changes in the pharmacokinetics of one or both of the drugs, and ultimately this may lead to either reduced efficacy and/or ADR. Veldkamp *et al.*, determined the pharmacokinetics of efavirenz or nevirapine over 24 hours when administered singly to patients with HIV, after which these individuals were switched to a combination therapy of both drugs. Following 4 weeks of treatment the pharmacokinetics of these drugs was once again determined. Combination therapy resulted in a 22% decrease in exposure (area under the curve, AUC) of patients to efavirenz, with the minimum plasma concentration falling by 36%. Such a decrease is almost certainly due to the induction of CYP3A4 and CYP2B6 by these compounds, leading to a marked increase in the rate of metabolism of efavirenz, which could potentially lead to a decrease in efficacy of this compound. With the knowledge of an antagonistic interaction, it was possible to calculate a new dose regimen that would produce plasma levels of both drugs identical to those seen if they were given singly, thus ensuring full clinical efficacy of both.

Antiretroviral drugs are not only given in combination with other antiretroviral drugs, however, but are often used in combination with other medicines targeted at other aspects of the disease. In the case of HIV, one such class of chemicals is the antibiotics, as HIV sufferers are predisposed to opportunistic infections, and in particular tuberculosis. Rifampicin is a commonly used antibiotic, and thus could be selected to treat such opportunistic bacterial infections. However, as with nevirapine and efavirenz, rifampicin is metabolized by, and is an inducer of, CYP3A4; it is therefore not difficult to envisage that this drug may affect the pharmacokinetics of these NNTIs. Lopez-Cortes *et al.* (2002) studied a group of HIV sufferers on efavirenz treatment who, having developed tuberculosis, were prescribed rifampicin. They demonstrated that while no change in rifampicin pharmacokinetics was observed, the peak plasma concentration and AUC of efavirenz were both significantly decreased, by 24% and 22% respectively. This could result in an effective under-dosing of this compound, with therapeutic levels not being reached and, in patients over 50 kg in weight the plasma concentration of efavirenz was halved. In this case, it can be calculated that to return the pharmacokinetic characteristics of efavirenz to that observed prior to co-administration with rifampicin an increase in dose from 600 mg/day to 800 mg/day was required. While the minimum effective plasma concentration of efavirenz has not currently been established it would seem prudent to alter the dosing regime to achieve a plasma concentration that has previously been shown to be effective, rather than hope that any lowering of the concentration by co-administration of rifampicin does not result in a drop below the minimum effective plasma concentration, and resultant loss of therapeutic efficacy.

6.2.4 Non-predictable mixtures

While the previously discussed 'predictable' mixtures are, as the name suggests, relatively easy to foresee and therefore investigate, a potentially more serious type of interaction is those derived from non-predictable mixtures. These interactions occur when an individual is exposed to two compounds that previously had not been thought of as a combination. Such exposure infers that the routes of exposure are different and the combination opportunistic. In general such combinations are not revealed until an adverse drug reaction is identified, at which point the investigation as to the cause of this reaction identifies the 'combination'.

One example of such a combination is that of a number of clinically important drugs and grapefruit juice. Terfenadine, a non-sedative antihistamine drug prescribed for hay fever is generally considered safe, but in the early 1990s several cases of toxicity were seen during its use. The toxicity was evidenced by an increased cardiac repolarization time (QTc) and has been linked to at least 125 deaths (Honig *et al.*, 1996). Analysis of the common factors between these cases revealed ingestion of grapefruit juice as the major linking factor. Analysis of the constituents of grapefruit juice revealed two chemicals, bergamottin and its metabolite 6'7'-dihydroxybergamottin, were inhibitors of CYP3A4 activity within the intestine. Terfenadine is converted to the active carboxylate metabolite by CYP3A4 in the intestine. In individuals taking grapefruit juice the rate of this conversion was markedly decreased and absorption into the plasma was mainly of terfenadine and not terfenadine carboxylate; this then reached toxic concentrations and caused the observed cardiac problems (*Figure 6.1*).

Such an analysis is retrospective, with the interaction only being identified after an ADR had occurred. However, by inference it is now possible to say that any drug that undergoes significant first-pass metabolism in the intestine by

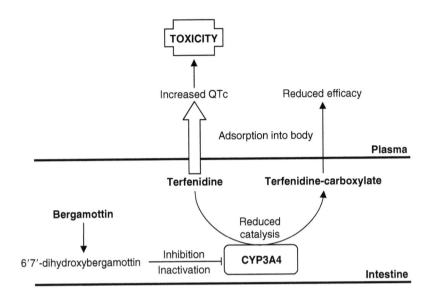

Figure 6.1

Mechanism-based inhibition of CYP3A4 by grapefruit juice constituents.

CYP3A4 may be subject to an interaction caused by grapefruit juice, and thus alter the risk assessments for these compounds appropriately.

A new source of potential interactions has arisen through the increased use of alternative herbal remedies, which often contain two of three active ingredients. If these are taken alongside traditional medicines then a simple mixture is created and the potential implications must be investigated. One of the most widespread herbal remedies used in the West at present is St John's Wort, an extract of *Hypericum perforatum*, used in main for the treatment of depression, but also suggested for everything from pulmonary complaints to worms.

Two chemicals are the major active constituents of St John's Wort, hyperforin and hypericin, the latter of which is thought to bestow the antidepressant activity of this mixture (*Figure 6.2*). In ligand-binding and reporter gene assays hyperforin, but not hypericin was shown to activate the steroid

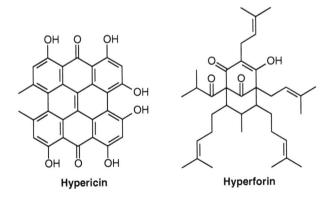

Figure 6.2

Active constituents of St John's wort.

hormone receptor PXR, and thus cause transcriptional activation of *CYP3A4* gene expression. This increased level of CYP3A4 results in altered pharmacokinetics of co-administered drugs and may therefore cause adverse effects. Indeed, a number of clinically important interactions have so far been identified, all of which result in reduced clinical efficacy of the medicinal drug (*Table 6.2*).

Table 6.2 Established interactions between St John's Wort and therapeutic medicines

Drug	Pharmacokinetic effect	Potential effect of interaction
Amitriptyline	↓AUC	Loss of antidepressant activity
Cyclosporin	↓Plasma levels	Organ graft rejection
Digoxin	↓AUC, ↓Cmax	No change in blood pressure
Indinavir	↓AUC, ↓Cmax	Loss of antiviral activity
Nevirapine	↑Clearance	Loss of antiviral activity
Oral contraceptives	None	Intermenstrual bleeding
Phenprocoumon	↓AUC	Loss of anticoagulant activity
Theophylline	↓Plasma levels	Loss of bronchodilator activity
Warfarin	None	Reduced anticoagulant activity

Adapted from Ioannides (2002).

6.3 Complex mixtures

6.3.1 Analysis methodologies

In the previous section we saw how it was possible to examine a simple mixture of less that ten chemicals, determine which components, if any, were interacting and what the overall effect of the combination would be. However, if you do not know what is in the mixture, either in terms of the actual components (qualitatively) or the concentrations of these components (quantitatively) how can you carry out such an assessment? This is often the problem with complex mixtures.

A complex mixture may contain thousands of chemicals in an unknown combination and hence different strategies must be employed to produce an accurate risk assessment for human exposure. Due to their inherent complexity, slightly differing assessment methodologies may be required for different mixtures, dependent upon the information present on the mixture: however, some general rules can be applied to analysis of complex mixtures.

This approach to risk assessment may depend upon the availability of the mixture; can the entire mixture be isolated for testing (available mixtures) or is it either a hypothetical mixture, or one that is not easily isolated (virtual mixtures).

If the mixture is 'available' and can be isolated in totality then one approach is to test the entire mixture and base risk assessment on this. While, as for testing of simple mixtures in entirety, such an approach will not identify interacting chemicals it will give an overall assessment of the mixture in terms of toxicological hazard. In addition, the inability to identify precisely which chemicals are interacting may not be an issue in a complex mixture as you may not know all the components anyway. However, if more information (either qualitatively or quantitatively) is known about the mixture then it may be possible to select the n most important chemicals, or chemical classes, and carry out a risk assessment of these as if they were a simple mixture. How though do you decide which are the most important? Such a decision can only be based upon weight of evidence, and may therefore be biased towards numerical, biological or toxicological importance depending upon the evidence available. One key factor however is that the definition of an 'important chemical' within a mixture is based upon the risk of that chemical and not its hazard. Because of this, you are selecting the chemicals that represent the greatest risk, as determined by a combination of their hazard and their prevalence within the mixture.

For 'virtual' combinations it is not possible to test the entire mixture, as it is not available. Hence, analysis of such complex mixtures can only be carried out by identifying the n most important chemicals in the mixture and examining those as a simple mixture.

In summary, risk assessment of complex mixtures is hindered by both the unknowns associated with such mixtures (what is in them and in what proportions) and the availability of the mixture (virtual versus available mixtures). The extent to which the former unknown can be addressed will be highly dependent upon the individual mixture, whereas the latter is a technical question to which no simple resolution is available. Whatever the individual factors for a mixture, it can be seen that for risk assessment purposes one major question must be answered – whether the mixture should/can be assessed in totality, or whether analysis of a simple mixture comprised of the most important chemicals in the mixture is more appropriate. To understand the application of such thought processes in the real world we shall examine two distinct complex mixtures, natural flavourings and drinking water treatment.

6.3.2 Natural flavouring complexes

The flavour and aroma of foodstuffs are due to a huge number of chemicals, the majority of which are of natural origin. Flavour is often enhanced by the addition of either artificial flavourings of increasing the amount of the natural flavourings present. Can we assess what effect adding extra flavouring to foodstuffs will have on the overall toxicological impact of that food? There are currently over 1700 natural substances that are recognized as flavour ingredients and hence the computation of potential interactions between them is an immense job. The chemicals are currently 'generally recognized as safe' (GRAS), based mainly on their natural presence in food anyway and low toxicity seen in animal studies. However, as the use of flavour additives increases re-examination of this status is required, due to potential new interactions caused by artificially altering the quality and quantity of chemicals present in foodstuffs. In 1993, the Flavour and Extract Manufacturers Association (FEMA) began a comprehensive review of these chemicals to ensure accurate risk assessment of their potential impact on human health. How then do you study such a large, diverse group of chemicals? Previously, we have seen how interactions are most common between chemicals with shared features, and the initial phase of this review was to subdivide these 1700+ chemicals into chemically related groups. Each of these groups is then assessed based upon the weight of evidence for each chemical within the group in terms of its potential interactions and toxicity profile and then an overall conclusion drawn. In many respects testing of dietary constituents presents a straightforward analysis as all animals are exposed to food mixtures every day. Adding a single chemical at a higher concentration into the diet alters its ratio within the mixture and lets you assess its impact towards the overall toxicity profile of the mixture, and indeed this was the process taken in assessing natural flavouring compounds.

Since the inception of the programme in 1993, five papers have been published on different chemical groups. The most recent paper focused on pyrazine derivatives, chemicals associated with roasted or toasted foodstuffs (Adams *et al.*, 2002). In addition to their natural levels in foodstuffs an estimated 2100 kg of pyrazine derivatives are used every year in the USA as flavour enhancers. While all of the 41 chemicals evaluated are naturally found in foodstuffs their concentration ranges over five orders of magnitude (from 0.001 ppm to 40 ppm). What then was the effect of adding extra of any one of these pyrazine derivatives into the diet of laboratory animals, and by extrapolation human diets? In general, toxicity was only evident when derivatives were added to the diet at levels many orders of magnitude higher than ever seen in commercial use. Hence, while the compounds do exhibit a low level of toxicity, their use within a mixture does not significantly increase the toxicity observed to the point where a real human health risk exists. This may not be surprising as the levels used commercially are also orders of magnitude lower than the levels naturally occurring in foodstuffs. On the basis of these data it was therefore possible to confirm the risk assessment that natural food flavouring pyrazine derivatives are 'generally recognized as safe'. This means that their continued use as food additives, at the current levels, does not pose a significant risk to human health.

6.3.3 Disinfection mixtures for water treatment

Water treatment involves the use of a number of hazardous chemicals such as chlorine or ozone to remove potentially harmful flora and fauna from water prior to public distribution. However, such treatment results in residues of these chemicals, plus by-products which remain in the treated water. While these chemicals are present within the water only at low concentrations, as human exposure to tap water is both universal and extends for the life of the individual it is important to assess what toxicological impact these residues may have. In addition we must assess if the complex mixture state they exist in causes any alteration in the disposition of these chemicals.

As stated earlier, a good method for studying complex mixtures is to identify the *n* most important chemicals in that mixture and then assess those as a simple chemical mixture. In drinking water the by-products of treatment are an important potential source of toxicity, and so the interactions of these must be studied. While such data are still under investigation, initial reports suggest that mixtures of these compounds do interact, but in a competitive fashion, resulting in a less than additive overall effect. As current risk assessment is based upon the presumption of an additive effect then it is likely this provides an over- rather than under-estimation of the risk associated with water disinfection (Driedger *et al.*, 2002).

6.4 Summary

While in general toxicology is thought of as the assessment of toxic damage caused by individual chemicals, it is important to remember that in real life no individual is exposed to such a simple situation. Instead we are constantly exposed to different mixtures of chemicals, some of which may interact and hence alter the overall effect they exert on our bodies. To determine the true risk to human health it is therefore necessary to examine these chemicals in the situation that they will be in when humans are exposed to them, that of a mixture.

References

Adams, T.B., Doull J., Feron, V.J., *et al.* (2002) The FEMA GRAS assessment of pyrazine derivatives used as flavour ingredients. *Food Chem. Toxicol.* **40**: 429–451.

Driedger, S.M., Eyles, J., Elliott, S.D., *et al.* (2002) Constructing scientific authorities: issue framing of chlorinated disinfection byproducts in public health. *Risk Anal.* **22**(4): 789–802.

EPA (1986) Guidelines for the health risk assessment of chemical mixtures. *Federal Register* **51**: 34014–34025.

Honig, P.K., Wortham, D.C., Lazarev, A., *et al.* (1996) Grapefruit juice alters the systemic bioavailability and cardiac repolarization of terfenadine in poor metabolisers of terfenadine. *J. Clin. Pharmacol.* **36**(4): 435–451.

Ioannides, C. (2002) Pharmacokinetic interactions between herbal remedies and medicinal drugs. *Xenobiotica* **32**(6): 451–478.

Lopez-Cortes, L.F., Ruiz-Valderas, R., Viciana, A., *et al.* (2002) Pharmacokinetic interaction between efavirenz and rifampicin in HIV-infected patients with tuberculosis. *Clin. Pharmacokinetics* **41**(9): 681–690.

McDonnell, P.J. and Jacobs, M.R. (2002) Hospital admissions resulting from preventable adverse drug reactions. *Ann. Pharmacotherapy* **36**: 1331–1336.

Mumtaz, M.M. and Durkin, P.D. (1992) A weight of evidence approach for assessing interactions in chemical mixtures. *Toxicol. Industrial Health* **8**: 377–406.

Mumtaz, M.M., De Rosa, C.T., Groten, J.P., *et al.* (1998) Estimation of toxicity of chemical mixtures through modelling of chemical interactions. *Env. Health Perspectives* **106**(s6): 1353–1360.

Tajima, O., Schoen, E.D., Feron, V.J., *et al.* (2002) Statistically designed experiments in a tiered approach to screen mixtures of *Fusarium* mycotoxins for possible interactions. *Food Chem. Toxicol.* **40**: 685–695.

Van den Berg, M. (1998) Toxic equivalency factors (TEFs) for PCBs. *Environmental Health Perspectives* **106**: 775–792.

Van den Berg, M., Birnbaum, L., Bosved, A.T.C., *et al.* (1998) Toxic equivalency factors (TEFs) for PCBs, PCDDs, PCDFs for human and wildlife. *Env. Health Perspectives* **106**(12): 775–792.

Veldkamp, A.I., Harris, M., Montaner, J.S.G., *et al.* (2001) The steady state pharmacokinetics of efavirenz and nevirapine when used in combination in human immunodeficiency virus type-1 infected persons. *J. Infect. Dis.* **184**: 37–42.

Role of genetics in toxic response

7

7.1 Introduction

One of the key, emerging, challenges in molecular toxicology is the application of genomics to understanding toxicity and in the discovery of new therapeutic targets. Such an approach provides us with a greater understanding of biological processes, and how these change upon toxic insult to the body. This knowledge will allow us not only to understand specific toxicities, and hence develop early markers for them but also aid in the discovery of novel compounds which lack, or minimize the risk of adverse side effects.

Genomic studies fit into three basic categories. Firstly, there are those designed to examine the fundamental effects of toxicants on an organism, thus providing a greater understanding of the molecular mechanisms of the toxic response. Secondly, studies into the differences in response between model species and humans enable more accurate extrapolation of animal model data to the human situation. Finally, studies on the interindividual variation within the human population are beginning to explain the spectrum of response seen within a population upon exposure to a compound, and thus more accurately set safe exposure limits. Such investigations will allow the future tailoring of drug design and/or treatment regimes to individuals or sub-groups of the population whose genetic make-up would result in adverse responses to normal drug therapy. Such techniques are, as we will see, already in place for some treatments but as our knowledge grows so will the ability to use this knowledge in a productive fashion to ensure better individual response to drug treatment.

7.2 Mechanisms of genetic control

Control of gene expression exists at several levels, allowing precise control of expression depending upon the requirements of any particular situation. The first level of control is that which is the minimum required to produce gene expression, the factors controlling *basal gene expression*. These basal mechanisms only give rise to a low level of expression and to achieve higher levels of expression in specific tissues of an organism, or during specific stages of development, factors are required to allow *constitutive gene expression*. Finally, the expression of many genes is capable of being switched on/off in response to specific stimuli, thus responding to changes in the cell's environment. Such *reactionary gene expression* is obviously a key factor in response to chemicals, either increasing expression of genes associated with drug metabolism or with combating the effects of a toxic insult.

7.2.1 Basal gene expression

A simple definition of a gene is a coding region (that encodes the protein product) and the elements which control its expression. This latter region can be functionally and spatially divided into two compartments: the promoter and enhancer regions. Promoters are the regions of DNA immediately 5′ to the transcription unit (coding region plus additional untranslated region) and are the sole requirement for basal gene expression. In comparison, enhancers and locus control regions are spatially distinct from the promoters, often by tens of thousands of bases, and, in conjunction with the promoter, are responsible for controlling the higher levels of gene expression associated with constitutive and reactionary gene expression.

In all eukaryotes mRNA is transcribed from genes by the enzyme RNA polymerase II, and hence the minimum requirement (basal gene expression) is to recruit this enzyme to the beginning of the coding region and specify the direction for transcription to occur. In the majority of mammalian genes (greater than 90%) a sequence element called the TATAA-box is found approximately 25 bp upstream of the transcription start site. This element is recognized by the TATA-binding protein (TBP), a subunit of the basal transcription factor TFIID. Once TFIID has bound to the TATAA-box it can recruit a number of other factors, including RNA polymerase II, which it guides to the transcription start site, thus initiating transcription. A full description of the transcription complex assembly is beyond the scope of this text, as is control of basal gene expression in genes with promoters lacking TATAA-boxes, but the interested reader is directed towards texts by Lewin (2000) and Latchman (2001). *Figure 7.1* shows the basal gene expression of the *albumin* gene, a classical TATAA-box regulated gene.

7.2.2 Constitutive gene expression

All of the basal requirements for the initiation of transcription, as described in the previous section, are, by necessity, ubiquitous. How then is gene expression regulated to occur in either a tissue- or developmental-stage-specific manner? The answer is twofold: specific factors can either activate of repress the binding of the basal transcription complex to the promoter.

For the transcription complex to assemble and initiate transcription it must have free access to the DNA at the promoter (i.e. the DNA must be in an 'open' conformation). However, DNA does not usually exist in a naked

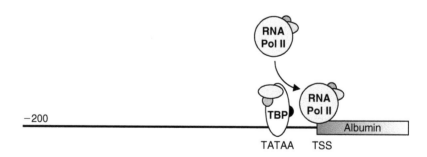

Figure 7.1

Elements involved in basal expression of the *Albumin* gene.

structure, and is instead packaged into a higher order chromatin structure. In this, DNA is first wound around an octamer of four different histone proteins to form a nucleosome. These nucleosomes are spaced along the DNA, forming a 'beads on a string' formation. This is then further coiled to produce tightly packaged chromatin, and in this state the transcription complex has no access to promoters (i.e. a 'closed' conformation) and so gene expression is repressed. To initiate gene expression, the chromatin structure must be opened to allow access for the transcription complex, and this is achieved in two ways. Firstly, histones can be covalently modified by histone acetyltransferases (HATs), adding acetyl groups to histone protein. As acetylation reduces the overall charge of the histone this will result in weakened interaction with DNA, and this was initially thought to be sufficient to cause a more open DNA structure. However, further work identified a second group of chromatin-modifying factors, and it is likely that acetylation of histones by HATs acts as a signal for this second set of enzymes, rather than dramatically altering chromatin structure directly. Chromatin remodelling complexes move histones along the DNA, without completely removing them, and thereby open the structure of regions of DNA. How then are these enzymes brought to bear on specific regions of DNA, the promoters of genes which are to be switched on?

Activation of transcription in a repressive chromatin environment is achieved by transcription factors. These proteins contain, at a minimum, a DNA binding domain which recognizes a specific DNA sequence, and an activation domain, which may work in one of a number of ways. In some cases the activation domain has either HAT or chromatin remodelling activity itself, and thus can directly improve access for the transcription factor to the DNA. Alternatively, the transcription factor may interact with HAT or chromatin remodelling complexes, recruiting them to the promoter, and so indirectly alter chromatin structure. Finally, activation domains may also interact with components of the basal transcription complex and so help recruit these essential factors to the promoter. Such recruitment may occur either through direct contact, or may require additional co-activator proteins to act as bridging molecules between the transcription factor and the basal transcription machinery.

Specificity of this regulation is achieved in two ways. Firstly, the DNA binding domain of a transcription factor recognizes a specific sequence of DNA (the response element) and hence transcription factors will only interact with promoters, and effect gene expression from promoters and enhancer that contain their specific response element. Secondly, the expression pattern of the transcription factor itself may be restricted to certain cell types, and hence these factors will only be active in those cells. It should be noted, however, that regulated gene expression is rarely under the control of a single transcription factor, and it is often the interplay between several different transcription factors that provide the precise regulation of gene expression.

An example of such tissue-specific expression is seen for the *Albumin* gene. The *Albumin* promoter contains binding sites for several transcription factors that interact to produce tissue-specific regulation of gene expression. Several response elements for *Hepatic Nuclear Factor* transcription factors (HNF-1) exist in the *Albumin* promoter. HNF1 is involved in chromatin remodelling of the *Albumin* promoter, thereby increasing access for other transcription factors such as C/EBPα, which in turn recruit the basal transcription machinery to the transcription start site. As HNF1 is expressed at much higher levels in the liver compared to other tissues, this means that the *Albumin* promoter exists in an open chromatin formation only in the liver, and therefore the *Albumin* gene is only expressed in this organ. *Figure 7.2* shows the promoter structure of the *Albumin* promoter, identifies the transcription factors responsible for its regulated gene expression, and shows how their tissue-specific expression results in the liver-specific expression of albumin (Lichtsteiner *et al.*, 1987).

As well as regulation through the promoter, genes also usually possess an additional regulatory region, the enhancer. Enhancers are physically distinct from the genes they regulate, often by thousand of bases, and can be either up- or downstream of the coding region; they are thus position and orientation independent. Like promoters, enhancers contain response elements for transcription factors involved in regulated or reactionary gene expression and often transcription factors that bind to a promoter of a gene also bind to the corresponding enhancer. When transcription factors bind to an enhancer they cause similar molecular events as seen with binding at the promoter (i.e. chromatin modification and recruitment of the basal transcription complex). The exact mode by which these spatially distinct enhancers interact with promoters is not known, although it is likely to be one of two mechanisms. While enhancers are many thousands of bases away from their promoters, they may be spatially very close, due to the 3-dimensional coiling of DNA in chromatin structure. Alternatively, binding of transcription factors to the enhancer may cause transmission of the signal down the DNA to the promoter, either through the alteration of chromatin structure, or through transmission of the signal via theoretical facilitator proteins. Whatever the exact mechanism of enhancer action, it can be seen that transcription factor binding to these regions acts to amplify the effects started at the promoter and thus further increase the rate of transcription.

Many genes are expressed not only in a tissue-specific pattern but also at a specific developmental stage. For example, in the human fetus CYP3A7 is the major liver CYP3A protein, whereas in the adult its expression is greatly reduced and CYP3A4 protein expression is switched on. How then

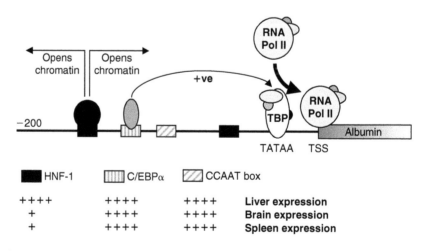

Figure 7.2

Elements involved in tissue-specific expression of the *Albumin* gene (adapted from Lichtsteiner *et al.*, 1987).

does the cell control this developmentally regulated gene expression? One simple explanation would be the developmental-stage-specific expression of transcription factors. This, as we saw in the case of tissue-specific expression, regulates expression of a gene to only the time when the crucial regulatory transcription factor(s) were expressed.

Perhaps the best studied developmentally regulated genes are the human β-globin genes, whose products are key components of the oxygen-carrying molecule haemoglobin. As can be seen from *Figure 7.3*, the human β-globin cluster extends over 70 kbp and consists of five genes, all of which show marked developmental regulation. Each gene has its own promoter but they share a single locus-control region (LCR), a specialized enhancer that co-ordinates the regulation of gene expression for the entire cluster.

Within the human β-globin cluster the genes are expressed in the order that they are found along the chromosome, and this phenomenon provided the first clues as to how these genes might be regulated through development. While each gene has its own promoter they must all compete for the LCR to enhance the efficiency of their promoters and initiate transcription. Experiments using transgenic mice shed further light on this regulation. Transgenic animals containing the start of the human cluster (LCR plus ε-gene) showed normal developmental regulation, fetal expression followed by decreasing expression with increasing post-natal age. However, replacing the ε-gene with the β-gene resulted in expression at all stages of development, showing no developmental regulation. Adding any gene between the LCR and the β-gene caused its expression to revert to the usual, adult-only expression profile. This suggested strongly that the developmental regulation was brought about by competition for the LCR, with genes nearer the LCR (i.e. ε) being expressed until their local, gene-specific, silencing elements switch them off. Only then can genes further from the LCR (i.e. β) be activated by it

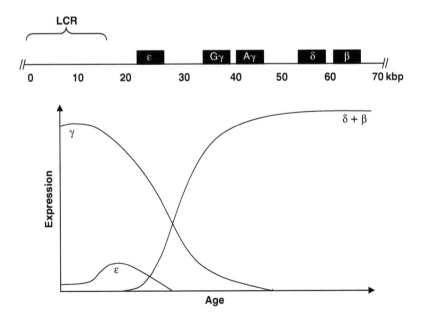

Figure 7.3

The human β-globin gene cluster.

(Enver *et al.*, 1990). It remains to be seen whether a similar mechanism is responsible for the developmental regulation of gene clusters involved in the response to toxic chemicals (i.e. CYP3A cluster).

7.2.3 Reactionary gene expression

Through the actions of tissue-, organ- and developmental-stage-specific transcription factors we can see how gene expression is controlled under normal physiological conditions. However, how does the cell alter gene expression in response to changing levels of a chemical? The transcription factors we have considered thus far are constitutively active and so their mere presence is enough to stimulate transcription. To provide an inducible control mechanism it is necessary to have transcription factors that are only active in the presence of a chemical, and these are termed ligand-activated transcription factors. The ligand in question may be endogenous or xenobiotic in source; it is merely the accumulation of it within the cell that causes the response. Ligand-activated receptors are generally present in the cytosol, where they are associated with chaperone molecules such as the heat shock proteins (HSPs). Binding of the ligand causes a conformational change in the receptor, resulting in disassociation from the chaperone molecules, translocation into the nucleus and exposure of the DNA-binding domain. These 'activated' factors may then bind to response elements within the promoters or enhancers of genes and cause an increase in the rate of the recruitment of the transcription complex to the promoter. The end result is an increase in the rate of transcription and so increased levels of the gene product designed to deal with a chemical insult, either through metabolism of the chemical or repair of any cellular damage caused by it.

7.2.4 CYP3A4: control of gene expression for a drug-metabolizing enzyme

As discussed in Chapter 3, the cytochrome P450 enzymes are responsible for the majority of Phase I metabolism, with CYP families CYP1-3 being involved in the primary metabolism of the majority of xenobiotics. It is hence not surprising that each of these families has a complex control mechanism to ensure gene expression is both constitutively present, but is also capable of responding to variations in concentration of the substrates for the enzymes. We shall look at the present understanding of how regulation of the *CYP3A4* gene occurs to show all the previously addressed elements integrating to form a functional regulatory unit for a gene's expression.

Figure 7.4 shows the basic structure of the CYP3A4 promoter and the interactions that occur to allow constitutive transcription of the gene. Binding of TBP to the TATAA-box and the subsequent recruitment of RNA polymerase II is enhanced by two regulatory transcription factors, SP1 and HNF-3. The former appears to aid in recruitment of the TBP, whereas the latter probably uses its HAT activity to alter chromatin structure into a more open structure that is more conducive to transcription. In addition, tissue-specific expression of HNF-3 confers tissue-specific expression of *CYP3A4* gene, predominantly in the liver and intestine. The overall result of this would be to produce a low level of expression of CYP3A4 within specific cells of the body.

As described above, to increase the level of gene expression and to allow reactionary gene expression many genes have response elements for

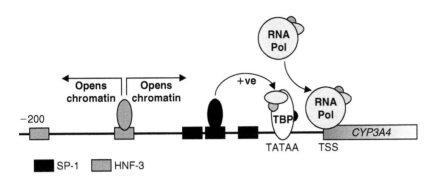

Figure 7.4

Elements involved in constitutive expression of the *CYP3A4* gene.

Table 7.1 Major ligand-activated receptors of CYP1-3 families

CYP	Receptor	Accession	Response element	Ligand
CYP1A	AhR	NM 001621.2	CTNGCGTGNGA	TCDD
CYP2B	CAR	NM 005122.1	TGTACTnnnnTGTACT	TCPOBOP
CYP3A	PXR	NM 003889.2	TGAACTnnnnnnAGTTCA	Rifampicin

Major ligand-activated receptors responsible for activation of hepatic CYPs are listed, along with their RefSeq accession number, consensus response element and example of a ligand. The *PXR* gene codes for multiple transcripts; only the RefSeq for the major transcript being indicated here. AhR = aryl hydrocarbon receptor, CAR = constitutive androstane receptor, PXR = pregnane X receptor.

ligand-activated receptors associated with them. For the CYPs, these transcription factors are activated by the same profile of chemicals as are substrates for the respective CYPs. *Table 7.1* shows the major ligand-activated receptors associated with the CYP1-3 families, along with the DNA sequence that they recognize. The presence of a pregnane X-receptor (PXR) response element at approximately −290 bp in the *CYP3A4* promoter provides the binding site for PXR, and confers the majority of xenobiotic-mediated gene regulation (*Figure 7.5*). In addition to allowing reactionary gene

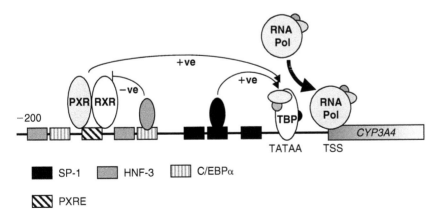

Figure 7.5

Elements involved in xenobiotic-mediated expression of the *CYP3A4* gene.

expression, PXR also inputs into regulated gene expression in two ways. Firstly, endogenous ligands for PXR will provide a low level of activation of this receptor under normal physiological conditions, and this will increase the constitutive expression of the *CYP3A4* gene. Secondly, tissue-specific expression of PXR helps to maintain the tissue-specific expression of CYP3A4, with high levels of PXR present in tissues such as liver and intestine where CYP3A4 is also expressed to high levels. Hence, interactions within the promoter provide all the necessary stimulus for tissue-specific and chemical-reactionary gene expression of CYP3A4.

While control of *CYP3A4* gene expression can be controlled just through interactions within the promoter, interactions within the CYP3A4 enhancer (termed xenobiotics-responsive enhancer module or XREM) fine tune this regulation and allow the amplification of reactionary gene expression effects. Goodwin *et al.* (1999) demonstrated the importance of enhancers in propagating response to xenobiotics through the examination of the *CYP3A4* XREM. Initially Goodwin *et al.* cloned over 13 kbp of DNA 5′ to the CYP3A4 coding region into a reporter gene. Then, through sequential deletion of sections of this clone they were able to identify 230 bp of DNA approximately 7.9 kbp away from the gene that was required to achieve maximal induction in response to xenobiotics. Bioinformatic and DNase I footprint analysis of this region identified binding sites for the Pregnane-X-receptor, as well as other, currently unidentified, transcription factors, within this region. PXR can therefore positively regulate expression of the *CYP3A4* gene when it binds to the XREM, as seen in *Figure 7.6.*

In summary, it can be seen that each of the levels of control of gene expression (basal, regulated and reactionary) are integrated to produce a

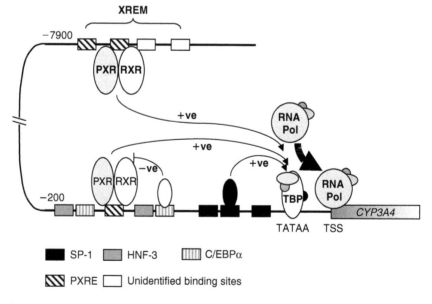

Figure 7.6

Enhancer contribution to xenobiotic-mediated control of *CYP3A4* gene expression.

powerful, precise control mechanism for gene expression. Through the use of specific transcription factors it is possible to provide both a tightly controlled expression (e.g. tissue or developmental expression) but with the ability to ramp-up expression in response to changing cellular circumstances (e.g. reactionary gene expression).

7.2.5 Co-ordination of response to toxic exposure

In general, exposure to a xenobiotic results in the activation of several genes, not just one. How then is this co-ordinated response achieved? The most usual way is through the presence of the same response elements in the promoters of several different genes. Thus, activation of a single ligand-activated receptor by a ligand will result in the activation of a number of different genes, whose increased expression will result in the co-ordinated cellular response to the chemical exposure. However, two other systems are often seen to co-ordinate the gene expression profile changes associated with exposure to a chemical.

In the first of these, many ligand-activated receptors exhibit a degree of overlap within the ligands that can bind and activate them. For example, glucocorticoids such as hydrocortisone are classically thought of as ligands for the glucocorticoid receptor (GRα). However, they are also capable of being ligands for the PXR receptor and it is the activation of both of these receptors that produces optimal induction of *CYP3A4* gene expression (El-Sankary *et al.*, 2000). Pascussi *et al.* (2000) demonstrated that this co-ordinated response was not just limited to binding of the receptors to the promoter and enhancer of *CYP3A4*. They demonstrated that glucocorticoids were capable of causing increased expression of the *PXR* gene, most probably through the binding of GRα to the *PXR* promoter. Hence, glucocorticoids not only activate PXR and GRα and cause them to bind to the CYP3A4 promoter/enhancer, but also increase the level of expression of PXR and its heterodimerization partner RXRα, thus further amplifying the induction response. Such co-ordinated induction of transcription factors extends beyond just ligand-activated transcription factors with, for example, glucocorticoids also known to increase the expression of the regulatory transcription factor C/EBPα (Grange *et al.*, 2001).

The other way in which a co-ordinated response can be achieved is through the sharing of response elements rather than ligands. Such co-ordination is exemplified by the ligand-activated receptors CAR and PXR. Maglich *et al.* (2002) examined the genes regulated by these two factors using receptor-selective agonists and receptor-null transgenic animals. They showed that while each receptor was involved in the regulation of several unique genes, a large subset of genes was regulated by both receptors, including a number of Phase I and Phase II drug-metabolizing enzymes. How then was such co-ordinated regulation achieved? The obvious solution would be the presence of response elements for both receptors within the promoters/enhancers of these genes. However, as demonstrated by Xie *et al.* (2000) this was *not* the case in this instance. Instead, Xie *et al.* demonstrated that CAR and PXR were capable of activating expression of CYPs through sharing response elements: CAR was therefore capable of binding to, and activating transcription from its native response element in the *CYP2B6* promoter and also from the PXR response element in the *CYP3A4* promoter, and vice versa for PXR.

This co-ordination of gene expression for genes involved in dealing with chemical exposures to the body is often termed the 'metabolic safety net' as it is thought to act as a survival system for the cell. If the primary system of drug metabolism is compromised in any way then the co-ordinated activation of alternate systems, either through activation of alternate receptors or one receptor activating the expression of alternate gene expression, helps to minimize the toxic insult placed upon the organism.

7.3 Tools for studying genetic responses to toxic insult

The advent of molecular biology has resulted in the development of a vast range of technologies for researchers to study the response of an organism to chemical exposure, be that a toxin or therapeutic agent, at the genetic level. Such technologies include those intended to study the overall response of an organism (e.g. DNA microarrays, differential display technology) to those which study smaller sets of genes (e.g. QRT-PCR, reporter gene assay, transgenic studies). An explanation of the underlying technologies is presented in Chapter 8, while examples of the use of these technologies in molecular toxicology are given below.

7.3.1 The study of whole genome responses to toxic insult

To fully understand an organism's response to a stimulus, be it therapeutic, environmental or toxicological, the ideal approach would be to measure the expression levels of every single gene before and after exposure. This is a daunting prospect considering that this corresponds to some 31 000 genes in most mammals. However, the advent of DNA microarray technology has allowed such approaches to become, at least to some extent, a reality. The spotting of several thousand gene fragments onto a single membrane means that a large portion of the genome can be studied at once. As it is not yet practical to fit the entire genome of a higher species onto a single chip, or to process such a large amount of data, many companies now produce 'themed' chips covering a particular area of interest (e.g. DNA synthesis and apoptosis or individual chromosomes). Interrogation of these microarrays with mRNA from exposed and naïve animals, and then data mining to distinguish genes which are expressed differently in the two situations, provides a wealth of information on how the genes represented on the microarray respond to the exposure. Such information can be used to either study the expression profile for a range of genes associated with particular cellular processes, perhaps to identify patterns of similar expression caused by a range of compounds, or to identify novel genes that are regulated via the compound. The latter approach provides an important extension of knowledge, as it allows the expansion of the known effects of a compound and how such effects may impact upon the observed final physiological effect.

We have previously examined the role of cytochrome b5 in reduction of chromium (Chapter 2) and have commented on the varied toxicities demonstrated by chromium-containing compounds, including cytotoxicity, mutagenicity and carcinogenicity. With such a wide range of toxic effects, it would be expected that a large number of genes is affected during chronic chromium exposure, either as a result of direct toxicity or due to the cell's attempts to repair the damage sustained. Such a complex toxic response lends itself to microarray analysis as such an approach allows the characterization of gene expression in many genes at one time. The overall aim of such an approach would be to understand the key cellular responses to chromium toxicity, plus the identification of potential markers for this type of toxicity. Cheng *et al.* (2002) used microarray analysis to investigate the effect of chromium(III) chloride on human testicular Sertoli cells. Following exposure of Sertoli cells to a non-toxic dose of chromium(III) chloride, mRNA was extracted and used to interrogate a microarray of approximately 10 000 sequences: 52 genes that showed changes in expression following exposure were identified. Of these, several coded for transcription factors and cell defence proteins and thus provided important information on the response of the cells to exposure to chromium(III) chloride.

Hamadeh *et al.* (2002) used microarray analysis to examine the changes in gene expression profile caused by rat exposure to three structurally distinct peroxisome proliferators (Wy-14,643, gemfibrozil and clofibrate) and compared them to the pattern caused by phenobarbital, which is not a peroxisome proliferator but does cause increased liver size similar to that caused by the peroxisome proliferators. Changes were observed in approximately 25% of genes on this 'toxicology chip', and two important distinctions could be made. Firstly, it was possible to discriminate between those genes whose expression changed only transiently and those for which the change was persistent. Transient changes tended to correlate with signalling molecules (e.g. G-protein coupled receptor kinase 5), whereas persistent changes were seen for genes encoding proteins involved in compound metabolism (e.g. CYPs) or cellular biochemical processes (e.g. histidine decarboxylase). Secondly, it was possible to mathematically separate phenobarbital from the peroxisome proliferators on the basis of the changes in gene expression observed. Further mining of this dataset will reveal which changes set these two classes of compounds apart and could then be used to understand the differences in molecular mechanisms underlying their actions.

An overview of the use of such DNA microarrays in the drug development process is provided by Gerhold *et al.* (2001).

7.3.2 The study of individual gene responses to toxic insult

The converse of studying global changes associated with a stimulus is to study the regulation of single pathways or even individual genes. Once the key pathways or enzymes involved in a toxic response have been identified, possibly through the use of microarray technologies, it is usual to concentrate on these key reactions to attempt to understand the molecular mechanisms underlying their genetic control. Such information can then be used to produce early markers for the associated toxicity, or to aid in the development of new compounds which do not exhibit these adverse effects.

A common method of studying transcriptional activation is the use of reporter gene assays, which link the transcriptional activation of the gene of interest with an easily measurable endpoint (e.g. secretory alkaline phosphatase; see Chapter 8 for details).

An example where reporter genes have been successfully employed is in the screening of compounds for their ability to induce cytochrome P450 enzymes. The wide substrate specificity of the CYPs produces the potential for problematic drug–drug interactions, with dosing of one compound increasing the levels of a CYP and therefore altering the pharmacokinetics of a second, co-administered, compound. This is particularly the case in the treatment of disorders where polypharmacy is prevalent (e.g. treatment of psychiatric disorders). Therefore, it is important to understand the induction profiles of compounds early in their development, in order to predict and therefore avoid potential drug–drug interaction problems once the compound reaches the general public.

Engineering a reporter gene so that it is under the control of a CYP promoter, coupled with transfection into human liver cells, allows the accurate assessment of a compound's ability to induce that CYP to be made. Using this technology it is therefore possible to predict the potential for drug–drug interactions to occur. El-Sankary et al. (2001) used a CYP3A4 reporter gene system to compare the induction profile of several clinically used compounds and were thus able to rank them according to their overall ability to induce CYP3A4 (Table 7.2). Such experiments provide important information on the possible implications of xenobiotic exposure on co-administered compounds and thus help to define safe conditions of usage for polypharmacy. The implications of exposure to mixtures of drugs are further discussed in Chapter 6.

An alternative use of reporter genes is for study of the transcription factors involved in the regulation of gene expression within a particular promoter. Through the use of site-directed mutagenesis it is possible to ablate specific response elements within a promoter and determine what effect this has on the expression of the reporter gene, either in terms of basal expression or response to chemical exposure. An example of such technology was provided by El-Sankary et al. (2002), who studied the effect of a mutation within the CYP3A4 promoter. Ablation of a putative C/EBPα binding site caused a decrease in the response of a reporter gene to glucocorticoids but not rifampicin, providing evidence that these two classes of compounds act, at least in part, through separate mechanisms.

7.4 Variation between species

Animal models have been used for many years to investigate compound effects and are a vital part of any data package for registration of medicinal drugs. Their use stems from the presumption that the response observed in other mammals is the same, or similar to, that observed in humans. Such a presumption is not without some basis, as many classical toxicants (e.g. TCDD, CCl_4) having very similar effects in rodent models compared to humans, causing overt toxicity of a similar pathology.

Table 7.2 Transcriptional activators of CYP3A4

Chemical	Clinical use	Inductive ability
Phenytoin	Anticonvulsant	Weak
Sulfinpyrazole	Uricosuric	Weak
Phenobarbital	Anticonvulsant/sedative	Weak
Clotrimazole	Broad spectrum antifungal	Medium
Fexofenadine	Hypersensitivity treatment	Medium
Rifampicin	Macrolide antibiotic	Medium/potent
Troglitazone	Antihyperglycaemic	Potent
Lovastatin	Cholesterol lowering	Potent

This may not be surprising when you consider how genetically similar we are to other mammals, a fact confirmed by the sequencing of the complete genomes of several species, including mice and humans. For example, mice and humans have a very similar number of genes (around 31 000) and most gene products in man have murine counterparts. In addition, many of the genes are arrayed along the chromosomes in the same order in the two species (synteny). Indeed, the X chromosome of humans and mice shows almost complete synteny, further emphasizing how genetically similar the two species are. Does such a similarity allow us to state that rodent models are the perfect system to study toxic responses and then to extrapolate this information to human exposure? Unfortunately, the answer must be no, for whilst there are striking similarities between the species there are also important differences (*Table 7.3*). To further highlight the differences and similarities we will examine two case studies where marked species differences exist; peroxisome proliferator-induced hepatocarcinogenesis and paracetamol-induced hepatotoxicity.

Table 7.3   Examples of species differences in toxic response

Compound	Responsive species	Mild or non-responsive	Humans response
Clofibrate	Rat	Rabbit	None
Paracetamol	Mouse	Rat	Mild to severe
Trichloroethylene	Mouse	Rat	Mild; not lung
Efavirenz	Rat	Cyn. monkey	None
Comfrey	Cattle	Sheep	Low (?)

7.4.1   Species differences in response to peroxisome proliferators

The peroxisome proliferators (PP) are a diverse group of compounds used in a number of industrial, domestic and medical situations (*Table 7.4*). Such usage means that human exposure can be high, or occur over a protracted period, emphasizing the need for accurate risk assessment for this class of compounds. PPs are negative in all standard genotoxicity assays, yet cause liver cancer in rats and mice during 2-year carcinogenicity studies – they are hence *non-genotoxic* hepatocarcinogens. With such a seemingly hazardous profile why are these compounds in such widespread human usage, and in situations where long-term exposure is a real possibility? The answer is that humans and mice/rats appear to respond differently to PPs.

Table 7.4   Varied uses of peroxisome proliferators

Chemical	Use
Gemfibrozil	Fibrate hypolipidaemic agent
Wy-14,643	Non-fibrate hypolipidaemic agent
DEHP	Plasticizer
Dicamba	Herbicide

DEHP = di(2-ethyl-hexyl)phthalate; Dicamba = 2-methoxy-3,6-dichlorobenzoic acid

Upon exposure to a PP rats and mice rapidly undergo an amazing transformation; within 48 hours the liver will have nearly doubled in size. This is caused by both an increase in hepatocyte number (hyperplasia) and an increase in the size of individual hepatocytes (hypertrophy). The latter phenomenon is caused by a large increase in the size and number of the subcellular organelle, the peroxisome, the event which provides this class of compounds with its name. Other species such as hamsters show some degree of hepatomegaly, while guinea-pigs appear to be refractory to the adverse effects of PPs. To assess the potential risk of PPs to humans it was therefore necessary to decide whether humans responded in a similar way to the sensitive rats, semi-responsive hamsters or apparently refractory guinea-pigs. To achieve this, it is necessary to understand the molecular mechanisms underlying peroxisome proliferation.

As discussed earlier, reactionary gene expression caused by exposure to a chemical is often mediated by ligand-activated transcription factors, and it was logical to presume that for peroxisome proliferation this was no different. The identification of the Peroxisome Proliferator Activated Receptor alpha (PPARα), for which PPs are ligands, therefore represented an important step forward in understanding the molecular mechanisms of peroxisome proliferation. However, just because peroxisome proliferators are ligands for PPARα it does not necessarily follow that PPARα is central to the hepatocarcinogenic response. How then do you prove such a link?

Two lines of evidence provided the evidence for the central role of PPARα in peroxisome proliferation. Firstly, cloning and sequencing of the promoters of several genes whose products are induced during peroxisome proliferation showed that they contained response elements for PPARα, termed PPREs. Using reporter genes it was possible to show that the PPRE was responsible for the observed induction response, as mutation of PPRE to prevent binding of PPARα resulted in a loss of transcriptional activation of the reporter gene in response to exposure to peroxisome proliferators. Secondly, transgenic mice were engineered to lack PPARα, and the response to peroxisome proliferators studied. Upon exposure of PPARα-null mice to the PPs clofibrate or Wy-14,643 none of the toxic effects usually associated with these compounds were seen; no increase in liver size or number of peroxisomes and most importantly, no liver cancer (Lee *et al.*, 1995).

Now that the molecular mechanisms underlying peroxisome proliferation had been linked to a single receptor it was much easier to study which animal model most closely resembles humans. The hamster, guinea-pig and human orthologues of PPARα were cloned and shown to be activated by PPs. In addition, the genes containing PPREs in rats and mice also contained them in these other species. What then is the difference between the responsive rodents and non-responsive guinea-pigs? While the receptors can be activated in both rats and guinea-pigs the number of receptors present in the liver varies greatly. Guinea-pigs express much lower levels of PPARα in the liver than rats and mice and this may explain their lack of response. As humans also express very low levels of hepatic PPARα it is plausible to suggest that guinea-pigs are the better model for the effect of PPs on humans, and risk assessment can therefore be made as an extrapolation of the guinea-pig, rather than rat, response to PPs (Cattley *et al.*, 1998). Hence, it appears that any PP will be safe in humans, provided that it is not a significantly better activator of human PPARα than its predecessors.

Thus the use of molecular methods has allowed the determination of a molecular mechanism, and such information is of use in the accurate risk assessment of the conditions under which new PPs may become a real risk to human health. The species differences observed in peroxisome proliferation are well reviewed by Choudhury *et al.* (2000).

7.4.2 Species differences in paracetamol toxicity

As discussed in Chapter 2, paracetamol, a normally safe analgesic, may cause severe hepatotoxicity in man under certain circumstances. Phase I metabolism by CYP2E1 results in the production of the highly reactive intermediate N-acetyl-p-benzoquinoneimine (NAPQI). Under normal physiological conditions NAPQI is quickly removed by Phase II metabolism, involving conjugation to glucoronide or sulphur. However, in situations where CYP2E1 is induced (e.g. following alcohol intake) or glucoronide depleted (e.g. following prolonged exposure to paracetamol) NAPQI accumulates and toxicity results. While this mechanism is now well understood, initial attempts to delineate the molecular mechanism were confounded by a marked species variation in the severity of response to paracetamol (*Table 7.3*).

Rats and rabbits are both insensitive to paracetamol toxicity, yet possess orthologues of all the genes which are responsible for the metabolism and toxicity of paracetamol in mice, which are sensitive to the toxicity. Why then do they not exhibit toxicity as well? The answer lies not in the mechanism of production of the toxic metabolite, but in the overall balance of the elimination pathways. Paracetamol only causes hepatotoxicity in mice when the rate of formation of NAPQI exceeds the liver's ability to remove it: hence, it is the *relative* and not the *absolute* levels of these enzymes that is important. *Table 7.5* shows the varying proportions of the Phase II metabolites excreted in different animal models. It can clearly be seen that mice and hamsters, species particularly sensitive to paracetamol-mediated hepatotoxicity, excrete a higher proportion of toxication pathway-related metabolites (i.e. glutathione conjugates and mercapturic acids) compared to less-sensitive species such as rats and rabbits.

Such data strongly suggested that species-specific effects are mediated by the levels of the enzymes responsible for paracetamol toxicity, or more specifically the ratio of those that lead to detoxification as opposed to toxication of the parent compound. The mode of paracetamol toxicity and the species differences in sensitivity is well reviewed by Bessems and Vermeulen (2001).

Table 7.5 Species-specific elimination of paracetamol metabolites

| Species | Sensitivity | % metabolites excreted | |
		Detoxification routes (GLUC & SULP)	Toxication route (Glutathione)
Mice	High	12	27–42
Hamster	High	41	27–42
Rat	Medium	62	5–7
Rabbit	Medium	27	5–7
Guinea-pig	Medium	74	5–7

7.5 Variation within a species

In the previous section the differences between species were highlighted along with the issues these raise in terms of extrapolation of toxicity data gained from model animals to the human situation. However, even if a suitable animal model is found there exists a potential for unpredicted human responses to occur. The deviation from the expected response can vary from relatively mild, subtly altering the overall efficacy of a compound, to large changes that see compounds either fail to exhibit a desired effect or cause an adverse response in the individual. This variation may be due to either sex differences or general population differences.

7.5.1 Sex-specific responses to toxic insult

If we revisit the definition of metabolism given in Chapter 1 we see that metabolism describes 'the total of all chemical transformations of normal body constituents'. As the 'normal body constituents' are different between sexes, particularly the levels of hormones, it is perhaps not surprising that metabolism may be different between males and females of the same species.

In general, sex-specific effects are more pronounced in rodent models than in humans, probably because the CYP enzymes that demonstrate marked sex-specificity (i.e. CYP2C11/12) in rats do not have human orthologues. Despite this, there are several cases of clinical drugs that exhibit sex-specific pharmacokinetic properties, as detailed in *Table 7.6*.

7.5.2 Population variation in response to toxic insult

As discussed in Chapter 6, humans are exposed to thousands of chemicals every day. Each of these has the potential to interact with other chemicals and alter the effect that chemical has on the body. Such effects are primarily through altering the disposition of drug-metabolizing enzymes within the body and thus changing the rate or route of metabolism of co-administered compounds. Should we expect a smoking, alcohol-drinking carnivore to react

Table 7.6   Sex-specific metabolism in humans

Chemical	Clinical use	Sex-specific effect
Aspirin	Analgesic, anticoagulant	Lower plasma levels in males
Chloramphenicol	Antibiotic	Lower plasma levels in males
Diazepam	Anxiolytic, anticonvulsant	Lower clearance in females
Erythromycin	Antibiotic	Lower clearance in males
Lidocaine	Anaesthetic	Increased $T\frac{1}{2}$ and V_D in females
Oxazepam	Anxiolytic, anticonvulsant	Lower clearance in females
Paracetamol	Analgesic, antipyretic	Lower plasma levels in females
Phenytoin	Anticonvulsant	Lower plasma levels in males
Propranolol	Antihypertensive	Lower clearance in females
Rifampicin	Antibiotic	Lower plasma levels in males

Adapted from Mugford and Kedderis (1998).

to all drugs in the same way as a non-smoking, tee-total vegan? Probably not. We must therefore take into account these differences when we make risk assessments on compounds, and in setting safe dose ranges.

However, even after eliminating sex-specific and environmental effects, a population will still show a range of responses to any one xenobiotic: Why? We have already seen that genetic differences between man and rodent can dramatically alter their overall response to xenobiotics. Small genetic differences may also exist *within* a species; any change in the DNA sequence away from the normal 'wild-type' is termed a *mutation*, and if this lies within a gene it may affect the gene product. Mutations in the coding region of a gene may cause altered protein activity (usually decreased), whereas mutations within the promoter or enhancer can affect the amount of normally active protein product produced. Thus mutations in the gene's drug-metabolizing enzymes, or proteins involved in response to toxic insult, may considerably affect the ability of an individual to respond to such a chemical exposure.

Mutation in a single individual, however, is not a major problem in terms of the whole population's response to a chemical insult. However, if the mutation occurs in the germ line it will be passed on to any offspring and can become fixed in the population. Once a mutation reaches a level of 1% in the population it is termed a *polymorphism*, and it is these differences that can cause significant variation in the overall response of a population to a chemical.

Chapters 2 and 3 contain brief overviews of the polymorphisms present in the human population for Phase I and Phase II drug-metabolizing enzymes, but the *CYP2D6* polymorphisms will be studied in more detail here. Variation in CYP2D6 activity was first identified though varied clinical response to the hypotensive agent debrisoquine. For the majority of people debrisoquine is the first-choice drug in the treatment of hypertension, however in a small number of people exposure leads to excessive hypotension. This adverse effect was shown to be due to the rate of 7-hydroxylation of debrisoquine, people with low rates of 7-hydroxylation accruing excess parent compound until toxic levels were reached (Gonzalez *et al.*, 1988). Two years later the oxidation of the alkaloid sparteine was also shown to be polymorphic, and the metabolic ratios ([parent drug]:[metabolite]) of these two chemicals was highly correlated in individuals, suggesting a shared route of metabolism. Further research identified CYP2D6 as the Phase I enzyme responsible for the initial metabolism of both of these compounds. Population analysis showed that CYP2D6 showed a bimodal distribution, with poor and extensive metabolizers, the former being susceptible to adverse drug effects (*Figure 7.7*). In European populations the frequency of poor metabolizers is approximately 7–10%, thus representing a small but significant portion of the populous. Genetic analysis of the *CYP2D6* gene has identified multiple alleles (currently >70), which cause increased, reduced or total ablation of CYP2D6 activity. Examples of the major alleles are shown in *Table 7.7*.

Another important factor about these, and indeed all, polymorphisms is that they tend to be geographically clustered, with some mutations being more prevalent in one geographical location than another. This is almost certainly due to the mechanism of formation of polymorphism; mutation of an individual followed by fixation into the breeding population. As people tend to breed with their neighbours, the spread of polymorphisms is centred on a single geographical location. This raises important questions for setting treatment regimes in different parts of the world. For example, in

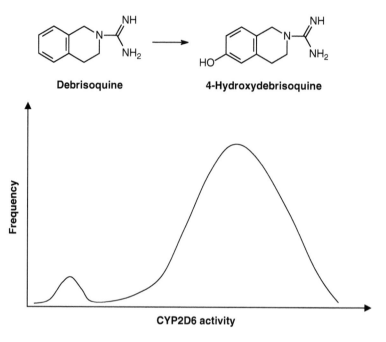

Figure 7.7

Metabolism of debrisoquine.

Table 7.7 it can be seen that 51% of the Chinese population has the CYP2D6*10 polymorphism, compared to less than 10% in other populations. As this polymorphism results in an unstable enzyme with reduced activity, it is easy to see how the Chinese population may have a higher incidence of adverse drug effects for CYP2D6 metabolized drugs than, for example, the Swedish population. Such data must be taken into account when deciding the most appropriate treatment regimes for different populations.

7.6 Summary

In this chapter we have examined how variation at the level of DNA sequences can result in markedly different responses following exposure to chemicals. Such variation may have its influence at one of two levels.

Table 7.7 Major alleles of the *CYP2D6* gene

Allele	Effect on protein	Allele frequency (%)		
		Swedes	Chinese	Zimbabwean
CYP2D6*1/*2	Wild-type	69	43	54
CYP2D6*3	Non-functional	2	0	0
CYP2D6*4	Non-functional	22	0–1	2
CYP2D6*5	No protein	4	6	4
CYP2D6*10	Unstable enzyme	n.d.	51	6
CYP2D6*17	Reduced affinity	n.d.	n.d.	34

n.d. = not determined. Adapted from Bertilsson *et al.* (2002).

Firstly, it may affect the regulatory regions of a gene, thus altering the level of either constitutive expression of the gene or the degree of reactionary gene expression in response to chemical stimuli. Alternatively, variation may exist within the region of a gene that encodes the protein product itself, and this can result in a protein product of altered activity. Therefore we need to understand this variation if we are to be able to fully understand individual responses to chemical stimuli.

The aforementioned variation exists both *between* species and *within* species. In the former case this is due to the process of evolution, and is what distinguishes two species apart. Such variation is, however, of great importance if we wish to use animal models of human behaviour. As we saw in the case of the peroxisome proliferators, one species (rat in this case) may be an inappropriate model due to its genetic make-up, whereas a second (i.e. hamster) may be more applicable. We must therefore consider this fully before extrapolating animal data to the human situation. In the second case, variation within a species, we must consider polymorphisms, and how these multiple versions of a gene may affect a single person's response to chemical stimuli. Many cases exist in the literature of marked interindividual variation in response to chemicals, and a large degree of this appears to be due to the mix of alleles that an individual has for the genes encoding the proteins required to deal with chemical insult.

A challenge for the next decade is to take this base knowledge and apply it to the field of toxicology. Already examples exist where knowledge of polymorphisms has explained variation in response to chemical actions and this must now be expanded to investigate all such variation, categorizing it as being due to either environmental or genetic variation.

References

Bertilsson, L., Dahl, M.L., Dalen, P., *et al.* (2002) Molecular genetics of CYP2D6: clinical relevance with focus on psychotropic drugs. *B. J. Clin. Pharmacol.* **53**(2): 111–122.

Bessems, J. and Vermeulen, N. (2001) Paracetamol (acetaminophen)-induced toxicity: Molecular and biochemical mechanism, analogues and protective approaches. *Crit. Rev. Toxic.* **31**(1): 55–138.

Cattley, R.C., DuLuca, J., Elcombe, C., *et al.* (1998) Do peroxisome proliferating compounds pose a hepatocarcinogenic hazard to humans? *Reg. Toxicol. Pharmacol.* **27**(1 Pt 1): 47–60.

Cheng, R.Y.S., Alvord, W.G., Powell, D., *et al.* (2002) Microarray analysis of altered gene expression in the TM4 sertoli-like cell line exposed to chromium(III) chloride. *Rep. Toxicol.* **16**: 223–236.

Choudhury, A.I., Chahal, S., Bell, A.R., *et al.* (2000) Species differences in peroxisome proliferation; mechanisms and relevance. *Mut. Res.* **448**(2): 201–212.

El-Sankary, W., Plant, N. and Gibson, G. (2000) Regulation of the CYP3A4 gene by hydrocortisone and xenobiotics: role of the glucocorticoid and pregnane X receptors. *Drug Metab. Disp.* **28**(5): 493–496.

El-Sankary, W., Gibson, G.G., Ayrton, A., *et al.* (2001) Use of a reporter gene assay to predict and rank the potency and efficacy of CYP3A4 inducers. *Drug Metab. Disp.* **29**(11): 1499–1504.

El-Sankary, W., Bombail, V., Gibson, G.G., *et al.* (2002) Glucocorticoid-mediated induction of CYP3A4 is decreased by disruption of a protein:DNA interaction distinct from the pregnane X receptor response element. *Drug Metab. Disp.* **30**(9): 1029–1034.

Enver, T., Raich, N., Ebens, A.J., *et al.* (1990) Developmental regulation of human fetal-to-adult globin gene switching in transgenic mice. *Nature* **344**: 309–313.

Gerhold, D., Lu, M.G., Xu, J., *et al.* (2001) Monitoring expression of genes involved in drug metabolism and toxicology using DNA microarrays. *Physiol. Genom.* **5**(4): 161–170.

Gonzalez, F., Skoda, R.C., Kimura, S., *et al.* (1988) Characterisation of the common genetic defects in humans deficient in debrisoquine metabolism. *Nature* **331**: 442–446.

Goodwin, B., Hodgson, E., and Liddle, C. (1999) The orphan human pregnane X receptor mediates the transcriptional activation of CYP3A4 by rifampicin through a distal enhancer module. *Mol. Pharmacol.* **56**(6): 1329–1339.

Grange, T., Cappabianca, L., Flavin M., *et al.* (2001) *In vivo* analysis of the model tyrosine aminotransferase gene reveals sequential steps in glucocorticoid receptor action. *Oncogene* **20**: 3028–3038.

Hamadeh, H.K., Bushel, P.R., Jayadev, S., *et al.* (2002) Gene expression analysis reveals chemical-specific profiles. *Toxicol. Sci.* **67**(2): 219–231.

Latchman, D. (2001) *Gene Regulation,* 4th edn. Nelson Thornes, Cheltenham.

Lee, S.S.T., Pineau, T., Drago, J., *et al.* (1995) Targeted disruption of the α isoform of the peroxisome proliferator-activated receptor gene in mice results in abolishment of the plietropic effects of peroxisome proliferators. *Mol. Cell. Biol.* **15**(6): 3012–3022.

Lewin, B. (2000) *Genes VII*. Oxford University Press, Oxford.

Lichtsteiner, S., Wuarin, J. and Schibler, U. (1987) The interplay of DNA-binding proteins on the promoter of the mouse albumin gene. *Cell* **51**: 963–973.

Maglich, J.M., Stoltz, C.M., Goodwin, B., *et al.* (2002) Nuclear pregnane x receptor and constitutive androstane receptor regulate overlapping but distinct sets of genes involved in xenobiotic detoxification. *Mol. Pharmacol.* **62**(3): 638–646.

Mugford, C.A. and Kedderis, G.L. (1998) Sex-dependent metabolism of xenobiotics. *Drug Metab. Rev.* **30**(3): 441–498.

Pascussi, J.-M., Drocourt, L., Fabre, J.-M., *et al.* (2000) Dexamethasone induces pregnane X receptor and retinoid X receptor-a expression in human hepatocytes: Synergistic increase of CYP3A4 induction by pregnane X receptor. *Mol. Pharmacol.* **58**: 361–372.

Xie, W., Barwick, J.L., Simmon, C.M., *et al.* (2000) Reciprocal activation of xenobiotic response genes by nuclear receptors SXR/PXR and CAR. *Genes Devel.* **14**: 3014–3023.

Technologies for toxicity assessment

8

8.1 Introduction

In the past decade a whole raft of technological advances has been made, and many of the discoveries detailed within this book were made solely through the use of these modern techniques. It is important to note that these 'modern' molecular techniques do not replace the 'traditional' biochemical tools used for the past 100 years, but they are a new, powerful, set of methods for looking at a question and thus complement the more traditional technologies.

This chapter describes some of the wide range of technologies now available to the molecular toxicologist to examine his/her area of choice. These technologies may be split into what are often termed the 'omics', describing the area of biology that the technology investigates. Genomics is the study of DNA, of the genome itself, looking at its regulation and variation. Transcriptomics on the other hand is the study of the mRNA expression profile derived from gene transcription of the genome. Proteomics is the study of the protein products generated by translation of the transcriptome, but also includes study of the post-translational modifications that occur to these proteins. Finally, metabonomics is the study of the overall effect the body has on a chemical, giving an overview of the relative activities of enzymes within the body. By studying each of these specialist areas we can produce integrated pictures of cellular response to toxic insult, and from this hypothesize the molecular mechanisms that underlie this response.

This chapter is not intended to cover all the possible technologies that a scientist could use, nor will it give detailed descriptions on how to carry out these techniques. The interested reader is directed toward the large number of practical, technology-driven, textbooks available or the scientific literature to answer such queries. Instead, the aim of this chapter is to provide a brief overview of the current, state of the art, molecular technologies available to study the mechanisms of cellular response to toxic insult.

8.2 Genomics

The study of the genome is perhaps the most basic starting point for all investigations into molecular mechanisms. Genomic investigations may be separated into three, broad categories: polymorphisms, regulation and transgenics. In the first, we can study the variation in DNA between individuals, looking for inherited variations. Such information is vital to understand the mixed response that individuals give to stimuli. Secondly, study of the genome can be used to examine the roles of regulatory sections of DNA that control gene transcription. Reporter gene analysis is a powerful tool in investigating how DNA sequences determine the transcriptome profile caused by any stimuli. Finally, to study the function of an individual protein, one approach is to completely ablate this protein, or replace it with alternate protein by changing the DNA that encodes it – such is the role of transgenic animal technology.

8.2.1 Analysis of variation within the genome

It is obvious that each individual looks different, with the exception of identical twins. This outward difference is due to the unique nature of the DNA that makes us what we are. Differences in the genes that we express within our cells, as well as the composition of these individual genes are what make each of us an individual. However, an important extension of this is that such subtle differences will exist on the inside as well, with difference in our genomes changing the overall make-up of proteins within one person's body as compared to another. This variation may alter the way we react to any toxic insult, and hence the categorization of genetic variation is an important goal for all biologists, including the toxicologist. How then do you go about searching for mutations in the 3×10^9 bp of DNA that make up each individual? Two approaches exist – targeted screening of important genes or shotgun screening of the entire genome.

The simplest way of ascertaining if mutations exist in a gene of importance for the body's defence mechanisms is to screen this gene in a number of individuals, and a number of technologies exist to do this. One popular technique is single-strand conformational polymorphism (SSCP) analysis. In SSCP analysis, sections of DNA are amplified by PCR and then denatured to single-stranded DNA (ssDNA) by heating. As the solution cools the ssDNA will re-anneal back to double-stranded DNA (dsDNA) through the formation of interstrand hydrogen bonds; however, some will form *intra*-strand hydrogen bonds and form complex 3-dimensional structures. The shape of this structure is dependent upon the exact sequence of the DNA. Hence, if an individual has a mutation within that region of DNA, their ssDNA will form a different structure and this can be identified by separation on a non-denaturing polyacrylamide gel (*Figure 8.1*). Once an individual with a mutation has been identified, that region of DNA can be sequenced to characterize the mutation. SSCP therefore enables the rapid screening of large numbers of individuals across many regions of DNA. This means that the positions of mutations can be roughly identified prior to more detailed secondary analysis.

Following identification of a mutation by SSCP its frequency in the general population or ethnic sub-populations must be determined. If a mutation is present in >1% of the population then it is termed a polymorphism, and can be thought to affect a significant portion of the population. As well as ascertaining the frequency of the mutation, experiments must also be carried out to determine the functional effect, if any, of the change. If the polymorphism occurs in the coding region of a gene such experiments will require production of the protein and assay of its stability and functionality; in comparison, if the mutation resides in the regulatory region of a gene then a reporter gene assay can be used to study its effect (section 8.2.2). Obviously, only polymorphisms that result in a significant change in biological function

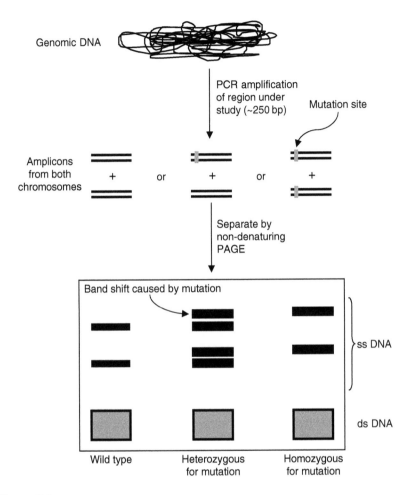

Figure 8.1

SSCP analysis to detect DNA mutations.

are of interest to the toxicologist, although it should be noted that 'silent' polymorphisms can act as important genetic markers, used by researchers in other areas of biology.

Another important part of genomic analysis is the ability to genotype individuals rapidly. This may be important in determining what alleles for key enzymes they possess, and hence how they are likely to respond to toxic insult. This may be important to understand why some individuals respond poorly to a compound, whether it is a therapeutic drug or environmental pollutant, while others do not. Genotyping of individuals can be easily achieved using two high-throughput technologies – 5' nuclease assay PCR and microarrays.

Section 8.3.2 will cover the use of the 5' nuclease assay to quantify how much of an mRNA transcript is expressed within a cell. However, this technology can also be adapted to provide the genotype of an individual. Probes specific for the different alleles are prepared, with each having a different fluorescent reporter dye. As this assay is so specific, even a single base-pair change can be detected, and the colour of the fluorescence will tell you which alleles an individual has. An alternative to this gene-by-gene approach is to use microarrays to genotype for multiple alleles on many genes at once. Section 8.3.1 covers the basic methodology of microarray technology, and it is easy to see how this can be altered to allow DNA screening; the attached target cDNAs merely target individual alleles rather than genes. As this screen looks at many more polymorphisms at once, the conditions may not be optimized for each hybridization, and hence it tends to produce more incorrect calls than the 5′ nuclease PCR approach; however, microarrays do allow a rapid snapshot of an individual's genotype to be gained. Microarrays are now available that contain the major polymorphic variants for a large number of toxicology-related genes, including Phase I and II enzymes, DNA synthesis and apoptosis-related genes.

8.2.2 Reporter gene assays

Reporter gene assays study the control of single genes at the transcriptional level and allow the assessment of whether changes in expression profiles are due to transcriptional activation, or by other means such as mRNA stabilization. In addition, they allow experiments to be undertaken to examine the molecular mechanisms that underlie the observed transcriptional activation.

As we saw in Chapter 7, expression of the coding region of a gene is under control of the promoter and enhancer of that gene; transcription factors bind to these and thus control the rate of RNA polymerase recruitment to the transcription start site. By cloning sections of promoters/enhancers into a vector containing the coding region of an easily measurable gene product, we can create a reporter gene that allows us to study these interactions (*Figure 8.2*). Easily measurable reporter genes include those producing enzymes that catalyse reactions with coloured endpoints (e.g. *lacZ*; the encoded enzyme converts the colourless X-gal to a blue metabolite) or luminescent endpoints (e.g. *Secretory alkaline phosphatase, SEAP*; the encoded enzyme can metabolize luminol, producing light in the process). Alternatively the gene product itself may be directly measurable, as with the fluorescent proteins (e.g. *Green fluorescent protein, GFP*). The amount of colour/luminescence or fluorescence is directly proportional to the amount of protein produced, and this in turn reflects the level of transcription caused by the DNA region under investigation.

Reporter gene systems are introduced into mammalian cells by the process of transfection; this will provide an environment containing all the transcription factors seen *in vivo*, producing the most accurate reproduction of the *in vivo* situation. The simplest method of transfection is transient transfection, whereby the introduction of plasmid is not permanent, and the reporter gene exists within the cell as isolated, circular plasmids. Addition of a compound that is a transcriptional inducer of the target gene produces increased levels of the reporter gene and this can be easily measured (*Figure 8.2*).

Such, transient, transfections have a number of limitations, due to the fact that the reporter gene is not maintained within the cells. Transfection of the reporter gene into the cells must therefore be done at the start of each experiment, and this is time consuming and costly. In addition, as the reporter gene is not maintained within a chromosomal environment, any higher DNA structures (i.e. chromatin) may not reflect the *in vivo* situation, and it may therefore respond in a different manner to the native gene within the genome. Because of these two factors alternative methods of transfections have been sought.

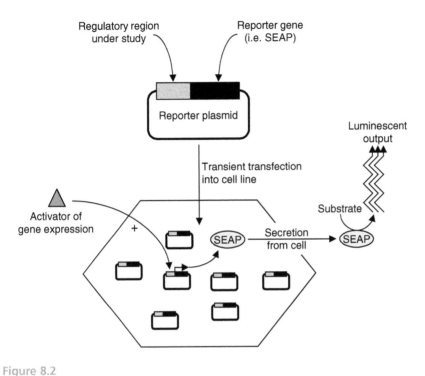

Figure 8.2

Reporter gene assays.

Rather than transient transfections, *stable* transfections can be carried out, where the reporter gene system is directly incorporated into the genome of the host cell. Therefore when the cell divides the DNA is passed into both daughter cells and will remain within the cells and their progeny in perpetuity. While the production of such stable cells requires a longer development period, with stable cell production taking from 2 months upwards, after this the cells can be used 'off the shelf'. Another advantage of stable transfections is that, in general, larger pieces of DNA can be incorporated (100 000 bp rather than a few thousand), and this means that more complex experiments examining both enhancers and promoters can be carried out. This may suggest that stable transfections are more useful that transient transfections, and while this is generally true they do suffer from two potential drawbacks; position-dependent variation and phenotypic instability of cell lines. A stated disadvantage of transient transfections is that as the reporter gene is not integrated into the genome then one level of control, chromatin structure, may be lost. However, in the construction of a stable line the site of integration may be such that the chromatin structure adversely affects the reporter gene functioning. For example, if a reporter gene integrates into an area of repressive heterochromatin then it will be poorly expressed, even if the promoter possesses strong transcriptional induction. Hence, it is usual to examine several different stable cell lines, with different integration sites, to conclude which is the most useful model to study. In addition, many cells in culture de-differentiate and take on a more fetal phenotype over time in culture. As they change phenotype their cohort of receptors and transcription factors may alter, and this in turn may alter the response of a reporter gene. Hence, to produce a valid, long-term stable cell line you must first ensure that the cells themselves are stable over time before even attempting to stably integrate the construct. *Table 8.1* lists some of the pros and cons of different reporter gene systems, comparing transient and stable constructs.

An example of the utility of reporter genes is the study of the regulation of CYP3A4 gene expression. El-Sankary *et al.* (2000, 2001) engineered 1 kb of CYP3A4 promoter upstream of a SEAP reporter gene and transiently transfected it into HepG2 cells (a human liver cell line). Using this system they examined the ability of various chemicals to induce activation of the reporter gene, studying both the relative strength of these activators and the roles of ligand-activated transcription factors in mediating this activation.

In addition to the use of reporter genes to merely assess transcriptional activation by xenobiotics, the molecular mechanisms of this activation may also be assessed, by measuring the effects of variants, either artificial mutants or natural polymorphisms, in the control regions of the gene of interest. El-Sankary *et al.* (2002) used an artificial mutant of the CYP3A4 reporter gene system described above to examine how transcription factors interact on the promoter. Using this technology they were able to identify transcription factor binding sites within the proximal promoter that are important in the activation of CYP3A4 transcription in the response to a variety of drugs. These sites have now been implicated in modulating the PXR-mediated activation of CYP3A4 gene expression by xenobiotics.

Table 8.1   Comparison of transient and stable transfection systems

Transient transfections	Stable transfections
No integration into host genome	Integration into host genome
Immediate set up	Two months plus to set up
Re-transfect for every experiment	Available 'off the shelf'
Does not model chromatin effects	Models chromatin effects
Control effect constant	Position-dependent control effects

8.2.3   Transgenics

Perhaps the ultimate method for testing the biological roles of a protein is to remove, or over-express that protein in a whole organism and observe the effects on the organism. Transgenic technology has allowed such powerful experiments to be undertaken, and has provided many new and exciting avenues of research. The basic principle of transgenic DNA technology is very simple, and relies on a basic biological phenomenon; homologous recombination. Within the nucleus of every cell we have two identical chromosomes, and if these line up next to each other it is possible for them to 'cross-over' and exchange sections of DNA. In general, this will have little biological effect, as the pieces of DNA are the same, but uneven recombination can result in new combinations of DNA and this can be important in both a positive way (progressing the evolution of DNA sequences) or a negative fashion (toxicity via clastogens). Transgenic technology uses this phenomenon to introduce DNA into targeted regions of genomic DNA, thus artificially changing the sequence. The simplest way of doing this is to target the centre of the gene of interest and introduce a region of DNA to disrupt the coding region of that gene; this produces a 'knock-out' that will lack the gene product of the targeted gene (*Figure 8.3*). Alternatively, an entire new gene (coding region and promoter) can be introduced into the genomic DNA, thus creating a 'knock-in' that will produce a gene product that it previously did not.

Using homologous recombination we can thus artificially introduce a section of DNA into genomic DNA, but how do we apply this to an entire organism. The most commonly used transgenic animal in toxicology research is the transgenic mouse and this is for two main reasons. Firstly, the responses of mice to many toxins are similar to human responses and hence they supply us with a good biological model, and, secondly, the technology for production of transgenic mice is well established. An overview of the transgenic process is given in *Figure 8.3*, and outlined in the text below. Embryonic stem cells (ES cells) are cells harvested from the blastocyst of a fertilized ovum.

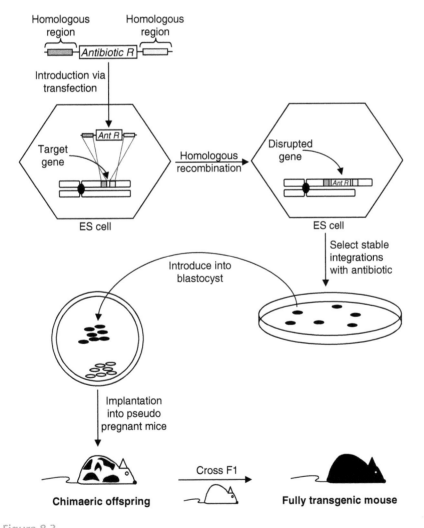

Figure 8.3

Production of a 'knock out' transgenic mouse.

They have two important properties that make them ideal for transgenic applications. Firstly, they are true stem cells and are capable of differentiating into any other cell type (i.e. totipotent). Secondly, ES cells from mice may be easily grown in culture, provided they are grown in de-differentiation medium to prevent them from beginning to differentiate. This means that ES cells need only be harvested once, thus reducing the need for animal sacrifice in the production of transgenics, an important ethical consideration. Once ES cells have been cultured it is a relatively simple procedure to transfect the section of target DNA into them, in a similar fashion to that used for making reporter genes (section 8.2.2). These constructs contain an antibiotic resistance gene, and the homologous DNA to the target area of the genome and any other DNA that one wishes to integrate (for example in a full gene in the case of knock-ins). Stable integration of the construct is monitored through the use of antibiotic selection over a period of 2–4 weeks; transient transfections, where the DNA target has not integrated into the genomic DNA, will lose the plasmid, and therefore the antibiotic resistance within this period and will then be killed by the antibiotic. Only cells with stable integration into the genomic DNA will survive such long-term antibiotic selection.

Once the target DNA has been stably integrated into the genomic DNA of the ES cells these cells can be introduced back into the blastocyst of a developing ovum, and this is used to fertilize a pseudo-pregnant female mouse. As this blastocyst will contain a mixture of the transgenic ES cells, and those already in there, the resultant offspring will be chimaeras: those sections of the animal derived from the transgenic ES cells will be transgenic and those derived from the normal ES cells will be wild-type. If some of the gonadal tissue is transgenic then cross-breeding of these animals will produce a pure transgenic animal that can be used for experiments (*Figure 8.3*).

8.3 Transcriptomics

Studies of genes and the various polymorphic alleles that they possess will provide us with much information on the *potential* responses to a toxic exposure. With this knowledge we can extrapolate what proteins can be made and hence hypothesize what the body *may* do in response to a toxic insult. However, to further understand the body's response to toxic exposure we must look beyond the genome to the mRNA produced, as this is an indicator of a cell's response to toxic stimuli. From this information we can gain an insight into the co-ordinated response of the body to the chemical exposure; such experiments are the remit of transcriptomics.

8.3.1 Microarray analysis

If you know very little about the biology of a cell's response to a toxin then perhaps the most obvious approach is to gain as much information as possible and then decide which are the interesting changes (i.e. causative of the physiological changes) and study these further – this is the approach adopted by microarray technology. Traditional techniques such as Northern blots have been used for many years to study the expression of an mRNA transcript. RNA from the tissue of interest is run on a gel, transferred to a filter and then hybridized with a labelled probe corresponding to the mRNA transcript of interest. Can this technology be expanded to allow the study of multiple transcripts at once? In microarray analysis the basic process is the same, hybridization of RNA to probes, but the overall *direction* of the reaction is opposite; specific target sequences are attached to the membrane and total RNA hybridized to this. RNA derived from an animal exposed to the toxin is labelled with one fluorophore, and RNA from a control animal is labelled with a second fluorophore. These labelled RNAs are both hybridized to the microarray, and the fluorescent colour of a spot will represent which RNA sample has more of the target. Using such an approach it is possible to produce a 'differential display' map of which mRNAs are present in higher, or lower, levels in one sample compared to the other (*Figure 8.4*). This information will

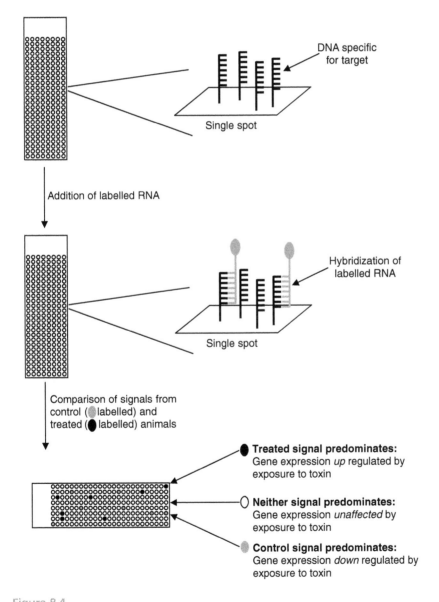

Figure 8.4

DNA microarray analysis.

thus tell you which genes have had their expression increased/decreased in response to the toxic insult, and thus give an overall picture of the response to the toxin. From this information hypotheses can be made and individual pathways chosen for further, more detailed, investigation.

The major advantage of microarray technology is that a large number of targets can be analysed at one time, with up to 20 000 individual spots being placed on a single microscope slide. These spots could represent 20 000 different targets, but this may raise the question of specificity – how can you be sure that each target is specific for its mRNA alone and not any closely related species? One way around this is to use multiple targets for each mRNA of interest and then look at the average signal from these. Many modern arrays, such as those sold by Affymetrix (http://www.affymetrix.com) now use such an approach, with 10–20 short targets for each mRNA of interest, and this is rapidly becoming the preferred method of analysis as the data gained appear to be more robust.

One development of particular interest to the toxicologist is the production of 'Tox chips'. These microarrays contain genes that are particularly relevant to the cell's response to toxic insult, and may also include multiple alleles for these genes. Such arrays allow the refinement of the initial question asked, and hence the data produced are biased towards toxicological responses and may be easier to interpret with respect to the molecular mechanisms of the cell's response to toxins. *Table 8.2* lists some of the gene groups commonly found on such 'Tox chips'.

8.3.2 Real time quantitative RT-PCR

While array technology will provide an approximate measure of the levels of many mRNAs within a sample, the hybridization kinetics of such complex mixtures mean that the number will not be fully quantitative. In addition, once you have identified the genes whose expression is altered following exposure to a chemical, further experiments should be focused on these rather than all the genes on an array. How then do you measure gene expression levels of selected genes in a fully quantitative manner? Any system designed to answer this question must show two characteristics – accurate, specific quantitation of the target and high throughput of samples. In meeting both these criteria we can develop a system that will allow the accurate

Table 8.2   Gene groups commonly used in toxicology microarrays

Gene group	Example genes/gene families
Phase I: CYPs	CYP1A1, CYP1B1, CYP2E1, CYP3A4
Phase I: FMOs	FMO3, FMO4, FMO5
Phase II: glucoronyl transferases	UGT1A6, UGT1A8, UGT2B7
Phase II: glutathione transferases	GST
Phase II: sulpho transferases	SULT1, SULT2
Drug transporters	OATP, MDR1, OAT, OCT1
Apoptosis	Bcl-2, Caspase-3, TNFα
DNA synthesis	PCNA
DNA repair	GADD153, GADD45
Oxidative stress	HSP70, cyctochrome c oxidase, MnSOD
Tumour suppressors	p53
Inflammation	Cox2, IL6

assessment of expression levels for a single gene product and will allow a high number of experimental points (i.e. dose and time) to be studied rapidly.

Polymerase chain reaction (PCR) experiments have led to a system that allows the rapid amplification of specific signals from small amounts of starting material. The addition of a reverse transcriptase step to first convert the unstable mRNA to more stable cDNA makes such technology applicable to examining mRNA levels. Amplification with gene-specific primers results in PCR amplification that occurs in four phases (*Figure 8.5*). Initially, amplification occurs below the detection limit, and hence no product can be detected (Phase I). Once amplification produces target over the detection limit an amplification phase is seen; during this phase the amount increases each cycle, doubling the amount of product each cycle if amplification is 100% efficient (Phase II). During Phase III, amplification moves away from exponential to linear, with some components of the reaction being limiting to the overall rate. Finally, one component of the reaction is used up and a plateau is reached and no more products produced regardless of increasing cycle number (Phase IV).

Traditional semi-quantitative PCR analysis relied on the determination of a cycle number mid-way along Phase II of the amplification curve. The *relative* amount of target in each sample is then assessed by comparing the amount of product produced by each sample at that cycle number. Such technology is *semi*-quantitative and not truly quantitative for two reasons. Firstly, no standard curve is used so expression is only ever calculated relative to the other samples. Secondly, due to the relative insensitivity of the agarose gel detection systems used to separate amplification products measurement must be undertaken relatively late in Phase III to ensure sufficient product is produced to be detected. This extended cycling will result in the amplification of any errors between samples (i.e. uneven loading), potentially producing inaccurate results. In addition, the fact that several optimization experiments must be undertaken to determine the cycle number midway along Phase II of the amplification curve means that such technology can be very time-consuming.

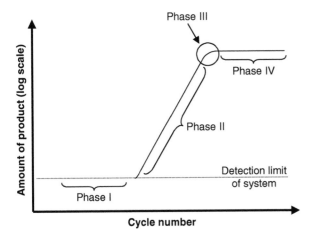

Figure 8.5

PCR amplification phases.

High throughput, quantitative PCR has been achieved using a number of technologies, but increasingly the most widely used is the fluorescent probe technologies employed in 5′ nuclease assays (TaqMan). In this technology, in addition to the two gene-specific primers a gene-specific probe is also used. By designing short amplicons (<100 bp) over half of the amplicon is covered by the two gene-specific primers (approximately 20 bp each) and the probe (approximately 15 bp). This results in an extremely high degree of specificity, allowing the accurate distinction between closely related mRNA species such as CYP3A4 and CYP3A5 (~90% identity in sequence along the entire mRNA), which would be very difficult to distinguish between using standard PCR.

The probe itself is comprised of a fluorescent reporter dye and a quencher separated by up to 15 bp of DNA specific to the target sequence. The reporter dye will emit fluorescent light when excited; however, when the probe is intact this energy is instead transferred to the quencher and no light is emitted. As can be seen from *Figure 8.6*, during each PCR cycle, amplification of the specific product results in the degradation of the probe, releasing the reporter from the quencher dye and thus allowing emission of light. Hence, fluorescent output is directly related to the cumulative amount of product produced. Fluorescent emission is measured at the end of each cycle of amplification, producing a plot which shows the full amplification profile for the reaction. The cycle number where the amplification amount of product crosses a threshold can then be measured (Ct or threshold cycle), and comparison of this value to a standard curve of genomic DNA will allow an accurate assessment of the *quantity* of target sequence in the original sample. As the fluorescent detection is highly sensitive Ct determination can occur very early in Phase II, markedly increasing the accuracy of results. Genomic DNA is used to construct the standard curve as we know the number of bases of DNA in a cell, and that this contains two copies of the target sequence, as each cell has two copies of each chromosome. We can hence calculate how many ng of DNA are required for 10, 100, 1000, etc., copies of the target.

As with all PCR technologies, these reactions may be carried out in a high-throughput methodology, using either 96- or 384-well plates. The addition of robotic systems means that assays may be run continuously, day and night. With an average run time of 1 hour and 40 minutes this means over 40 000 reactions can be completed per week.

One potential confounding factor in quantitative PCR is the decision of whether to use internal standards to ensure even loading of template material into each reaction tube or not. Internal standards are generally housekeeping genes such as actin, GAPDH, etc., that are expressed in all cells at high levels; if loading is even their levels should remain constant in all samples. While such an approach is valid in theory, extended literature has shown that levels of housekeeping genes do alter, both during the normal circadian rhythm, and also in response to chemical treatment. Hence, to validate the use of such internal standards it is important to show that they are invariant in your samples, or to use several standards, using their average expression to normalize the data. As both approaches are technically and scientifically complex to achieve, the most commonly used approach is to use an accurate method of template quantitation, such as a fluorescent dye, and quote levels as 'per ng input mRNA'.

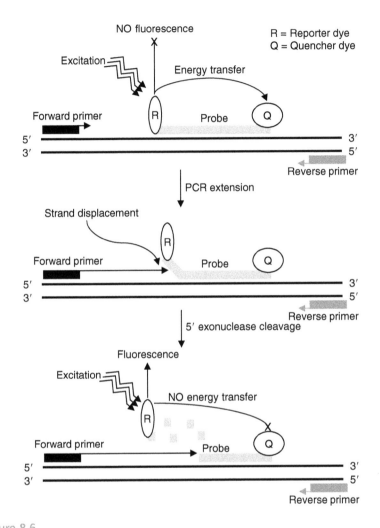

Figure 8.6

Quantitative PCR: the 5′ nuclease assay.

8.4 Proteomics

In the previous section we looked at the technologies designed for examining levels of mRNA transcripts within a cell. These technologies are now at a relatively high level of development and allow in-depth analysis of a cell's response to a chemical. If we can study the levels of transcripts so closely do we need to look at the levels of proteins? The answer to this must be an emphatic yes for a number of very good reasons. Firstly, proteins are the molecules that actually carry out the biological functions, whereas mRNA only codes for it. Hence, if we want to understand how the body works then we really must study the molecules that carry out these functions rather than extrapolating from their precursors, mRNA transcripts. In addition, it is becoming increasingly clear that mRNA changes do not always accurately reflect changes seen at the protein level. Translational efficiencies vary between mRNA transcripts and therefore 100 molecules of one mRNA transcript may result in 1000 proteins in one case but 10 000 proteins in another. Because of this, study of transcript levels may over- or underestimate the numerical abundance of a protein. Therefore, studying protein levels may give us a more relevant picture. Why then study transcriptome profiles at all? The answer to this is twofold. Firstly, study of transcriptomics allows us to understand the basic molecular mechanisms of cellular response to toxic insult, and how this is controlled at the level of transcription. The second reason is purely a technical one; the technologies for studying changes at the transcriptome level are more advanced than those for analysis of the proteome. While studying the levels of individual proteins is within the remit of most laboratories, the ability to undertake large-scale analysis of the same complexity, sensitivity and resolution as is currently achieved for the transcriptome is not possible, although the rate of technological advance means that it may soon be a reality. What then, currently, can we gain from proteomics studies?

8.4.1 2D-gel electrophoresis

Perhaps one of the oldest protein analysis methods is that of SDS-polyacrylamide gel electrophoresis (SDS-PAGE). In this technique, proteins are denatured and separated according to size through a polyacrylamide matrix. While such a method allows the separation of several proteins at once it does have one obvious limitation – how do you separate proteins of the same size? As the complexity of the protein mixture loaded onto a gel increases so does the chance of there being two proteins with approximately the same size. The number of proteins resolved on a single gel can be significantly increased if rather than separating the samples by a single dimension (i.e. by size) they are separated in *two* dimensions. In the case of 2D-gel electrophoresis this second dimension is provided by separation according to isoelectric point (pI). Every protein is comprised of amino acids and, according to the Henderson–Hasselbach equation (*Figure 8.7*), may be either charged or neutral. If a protein has a net charge (i.e. the sum of charges of all its constituent amino acids) then it will migrate through a gel under an electric current. However, at one pH the net charge of the protein will be zero and the protein will not migrate; this is the pI. Hence, proteins are separated through a pH gradient gel, and stop migrating once they reach the pH that corresponds to their pI. Following separation of proteins by pI point, separation by size is carried out at right angles to the first separation, and this will result in a 2D-gel where the resolution may be increased from <50 proteins (1-D electrophoresis) to up to 10 000 proteins (2D electrophoresis; *Figure 8.8*). An additional benefit of this approach is that the initial separation is dependent upon the IE of the protein, and this in turn is determined by the exact nature of the protein. Even the addition of a single phosphate group will alter the IE sufficiently to allow separation on a 2D gel. This therefore adds a powerful tool to the molecular toxicologist's armoury – the ability to monitor post-translational modifications. As covered

$$pH = pK + \log \frac{RCOO^-}{RCOOH}$$

Figure 8.7

Henderson–Hasselbach equation

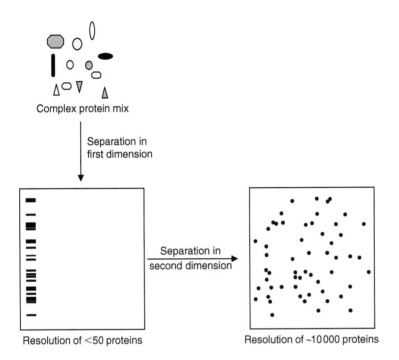

Complex protein mix

Separation in
first dimension

Separation in
second dimension

Resolution of <50 proteins

Resolution of ~10 000 proteins

Figure 8.8

1D vs 2D polyacrylamide gel electrophoresis.

in Chapter 4, many protein activities are regulated by post-translational events such as phosphorylation, and hence the presence of a protein product does not necessarily correlate with protein activity. With 2D-gel electrophoresis it is possible to monitor the presence of both phosphorylated and unphosphorylated proteins, and therefore to gain an insight into biological activation as well as mere presence.

The ability to resolve large numbers of proteins, and to distinguish between different post-translational modification states of these proteins, allowed differential display techniques, similar to those undertaken for the transcriptome, to be carried out on the proteome.

Despite the improvements in resolution afforded by separating proteins in two dimensions (size and pI), this technology is still limited in the number of proteins it can resolve (approximately 10 000) relative to the total number that may be present within the human body (probably in the range 200 000–300 000). Therefore, at the current time it is difficult to separate all of these proteins at once, and this means that any analysis will be incomplete. Thus, the next challenge for the protein biochemist is the refinement of technologies to allow the resolution of more proteins at any one time, allowing proteome-wide experiments to be undertaken in the same way as genome/transcriptome-wide analysis is currently available through microarray technology.

8.4.2 MALDI-TOF mass spectroscopy

2D PAGE has allowed the separation of large numbers of proteins, including the post-translational modifications (e.g. phosphorylation) of a single protein.

Using this technology we can therefore look for changes in the proteome caused by chemical treatment by comparing 2D PAGE from control animals and those exposed to the toxic compound. However, this will only identify new *spots* on the gel, to identify *what protein* these spots represent they must be excised from the gel and further analysed. The instrument most commonly used for this analysis is the matrix-assisted laser desorption ionization (MALDI) time-of-flight (TOF) machine. In such an analysis, the excised spots are digested with trypsin to reduce the proteins to small peptide fragments, which are attached to a matrix and placed in the MALDI-TOF machine. The fragments are excited with a laser causing ionization of them and then attracted towards a detector with a powerful magnet. The time-of-flight from the matrix to the detector is proportional to the size of the fragment, and hence a highly accurate mass can be gained. Such analysis results in a series of peaks, characteristic of the different peptides found within the digested spot. Bioinformatic analysis of the resultant pattern of peaks, using software tools such as MASCOT and PROWL, provides probability identifications for the protein spot against the SWISS-PROT database (http://ca.expasy.org/sprot/sprot-top.html), which contains the patterns of over 150 000 known proteins (*Figure 8.9*).

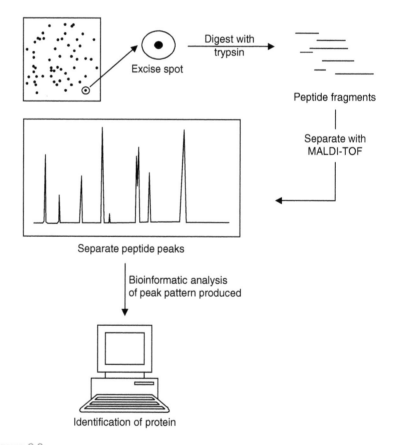

Figure 8.9

Protein identification using MALDI-TOF (adapted from Gygi and Aebersold 2000).

MALDI-TOF is not the only mass spectroscopy method used for such research, and quadrupole-time-of-flight (Q-TOF), fourier-transform ion cyclotron resonance mass spectroscopy (FT-ICR MS) and tandem mass spectroscopy (MS-MS) have all been applied to proteomic analysis. Each system has its own advantages and disadvantages, and while MALDI-TOF remains the most commonly used method other systems are being designed to answer specific questions. *Table 8.3* lists some of the instruments used, and their relative advantages.

An extension of this *descriptive* proteomics is *quantitative* proteomics. The obvious approach for this would be the use of a standard curve of known amounts of protein, in the same way as a genomic DNA was used to construct a standard curve for quantitative RT-PCR. However, as the ionization efficiency of different peptides is highly variable the only true reference for any peptide is that very peptide. Therefore, in quantitative proteomics, samples are 'spiked' with known amounts of internal standards of the same peptide labelled with a 'heavy' isotope (usually ^{2}H, ^{13}C or ^{15}N labelled). As the chemico-physical properties of the two peptides are the same, they progress through the MALDI-TOF in the same fashion, with only a small difference in mass, due to the labelled isotope. Quantitation of the sample peptide can therefore be assessed as a ratio to the internal standard.

Table 8.3   Mass spectroscopy instruments used for proteome analysis

Instrument	Characteristic
MALDI-TOF	Suited for analysis of less complex patterns
Q-TOF	High mass accuracy. High sensitivity
MALDI-Q-TOF	Increased throughput compared to normal Q-TOF
MALDI-TOF-TOF	Used for high energy collision induced disassociation (CID) studies
FT-ICR MS	Very high sensitivity and resolution

Data from Gygi and Aebersold (2000).

8.4.3   Protein chip analysis

A relatively recent development in the field of proteomics is the protein chip. This development stems from the field of transcriptome research; if we can develop high-throughput microarrays for DNA/RNA analysis could we do the same for protein analysis?

The basis for protein chips is a surface treatment to which target proteins will bind. This can be either affinity chromatography based (i.e. reverse phase, cationic, anionic or metal affinity) or biological (i.e. antibody, DNA, ligand affinity). Cell extracts are hybridized to the protein chip and then non-specific interactions removed through a series of washes. Finally, the captured proteins are analysed by surface-enhanced laser desorption/ionization mass spectrometry (SELDI), producing molecular mass information that can be used to identify the protein.

An example of the potential power of such protein chips was their use in the characterization of the α-defensin proteins (1, 2 and 3). Since 1986, it has been known that CD8 T lymphocytes from some HIV patients secreted a factor that suppressed HIV replication, and this stabilized the patient's immune system. However, this protein(s) had been extremely hard to characterize, and it was not until Zhang *et al.* (2002) utilized protein chip technology that the α-defensins were isolated. This discovery is a major step forward in understanding the body's defences against HIV infection.

8.5 Metabonomics

Metabonomics may be described as 'the determination of systemic bio-chemical profiles and regulation of function in whole organisms by analysing biofluids and tissues' (Nicholson *et al.* 2002). Such data should be of great value, as they are a direct measure of the biological consequences of chemical exposure; in comparison transcriptomics and proteomics study changes that may *ultimately* lead to these biological consequences. How then do you carry out such studies?

Similar technologies as that applied to proteomics (i.e. mass spectroscopy) could be used, but one technology, NMR has several advantages. NMR is non-destructive to the sample (MS requires digestion of the protein sample into fragments), can be used on unpurified biosamples (i.e. urine) and provides more information than MS with respect to the determination of molecular structures. How then does NMR work? NMR relies on the magnetic properties of certain atomic nuclei, and the fact that these properties are dependent upon the molecular environment of the atom. Atoms such as ^{1}H, ^{13}C, ^{31}P and ^{15}N have an odd mass number, meaning that they have an uneven number of protons and neutrons. This means that the atom has a magnetic moment (i.e. positive and negative sections of the atom). Application of a magnetic field can cause these sections to flip, causing a measurable energy fluctuation. NMR machines scan through energy emission in the MHz spectrum, and record fluctuations that represent molecular resonances. The resultant peak pattern can then be analysed to detect patterns associated with chemical exposure.

One application of metabonomics is the identification and detection of biomarkers. How do I tell if an individual has been exposed to a chemical? Measuring either the transcriptomic or proteomic changes known to be caused by exposure to this chemical would be a valid approach; however, to do this an invasive procedure would be required to gain the tissue to carry out such an analysis. A much better alternative would be to identify metabonomic changes that occur and then measure these in the urine, thus providing a non-invasive method of assessing if exposure to the chemical had occurred, and if so what biochemical effects this had caused within the body. For example, Beckwith-Hall *et al.* (1998) used biofluid NMR to study the metabonomic profile of urine from rats exposed to the three hepato-toxins α-naphthyl isothiocyanate, galactosamine and butylated hydroxy-toluene. They were able to distinguish the three hepatotoxins on the basis of their metabonomic profiles, as well as distinguishing between the biliary and parenchymal injury caused by these compounds. This produced, for the first time, specific biomarkers for exposure to these compounds, as well as indicators of the type of damage the body had sustained.

8.6 Bioinformatics

The advent of modern, high-throughput analysis of the 'omes' has produced a vast amount of information on how biological systems function, and their responses to toxic insult. One obvious way forward then is to fully analyse these data in an attempt to gain new insights from them, rather than to repeat experiments already done by somebody else. For example, two, apparently unrelated, experiments, when looked at together may yield vital information on the current area of study. Fortunately the development of the World Wide Web (WWW) has produced an amazing forum for scientists to exchange ideas and technologies, and to extract new information from published data. Unfortunately, the rapid expansion of the WWW has also resulted in a lack of cohesion in how these data are presented, and finding relevant information can be a major challenge: how then does one make the most of this resource?

In this section useful resources for the molecular toxicologist are listed by target area (i.e. genomics, proteomics, etc.), along with a description of how this information can be used to gain an overall picture of cellular responses to drug action.

8.6.1 Searching sequence databases

A piece of DNA/RNA is merely a string of nucleotides, and on that level yields little information. Thus, the first job once a novel piece of DNA/RNA has been isolated via genome or transcriptome analysis should be to gain more information about it, and an easy way to do this is via the WWW. Perhaps the first question is 'Has anybody else found this before?', followed rapidly by 'What does it do?' and 'In what other species is it found?'. Such questions are relatively easy to answer due to the collation of sequencing information in freely accessible databases on the WWW such as that maintained by the National Center for Biotechnology Information (NCBI; http://www.ncbi.nlm.nih.gov/).

As all the sequence data are now held on these central databases, tools are needed to search for relevant information. This can be achieved using one of three basic strategies.

1. The name of the gene of interest can be put in, plus any other key words such as the species of interest. This type of searching is only as good as the information you supply, the less precise the enquiry the more undesirable hits you will achieve.
2. A much more precise method is the use of accession numbers. Each sequence submitted to a database is given a unique accession number, and searching with this number will only bring up this sequence. However, due to the many sequencing projects around the world there may be several versions of the same stretch of DNA/RNA in the database – which one do you use? The RefSeq accession number refers to the gold-standard sequence for a particular stretch of DNA/RNA, and hence should be the one searched for if at all possible (and is used throughout this book).
3. Finally, the third approach applies if you don't have enough information to name the protein or quote its accession number, and that is searching with the sequence itself.

In 1990, Altschul *et al.* (1990) developed an algorithm for comparing a sequence to a database and coming up with matches, and this basic local algorithm search tool (BLAST) has become a key weapon in the bioinformatician's armoury. A BLAST search will interrogate the database and return matches to your target sequence, along with key information about each of these hits. Firstly, the matches are ranked in order, with the closest match to your sequence first, and their accession number and name given. These are hypertext linked so that you can jump to the database entry for that gene (via the accession number) or a graphical representation of the area of match between the two sequences (via the point score). Following this is information on how good the match is, expressed as a probability score. Values with $P < 0.01$ show a high degree of confidence that the match is real and not a result of chance. Due to the conservation of sequences through evolution such a search will also answer the third of our initial questions: 'What other species has this been found in?' Many of the hits generated from a BLAST search will represent orthologues from other species, and comparison of these to the target sequence will tell us how much this sequence has diverged over time.

Once you have located an entry for your sequence within the database, you may access this entry to gain further information about your sequence. Each entry in the database contains several bits of important information: basic information includes name, accession number, species, submission details and sequence. In addition, characteristics of the sequence are predicted; such characteristics include intron–exon boundaries and transcription/translation start sites for genomic DNA sequences and if the sequence encodes a protein then the predicted protein sequence is given.

With the rapid development of the sequence databases it became desirable to subdivide them into sections, thus allowing more precise searching. One worthy of special note is the Expressed Sequence Tag (EST) database. Not all sequences held on the database have a gene/function attributed to them. However, if the sequence was derived from mRNA then it is known that this sequence forms part of a transcript that encodes for a protein product, even if the role of that product is unknown. These are thus referred to as ESTs since they are known to be expressed. Searching of EST databases may provide important information on the target sequence, and is commonly used for two applications – identification of coding regions of mRNA transcripts and identification of putative polymorphisms.

mRNA transcripts are not made up solely of the open reading frame that codes for the result protein. All mRNA transcripts also contain untranslated regions (UTRs) before and after the open reading frame, and these are important in the stability and regulation of the mRNA. If your target sequence lies within one of these untranslated regions it may not produce a significant hit in the main database, for two reasons. Firstly, the main databases are biased towards coding regions of DNA and hence may not contain matching sequences to your target. Secondly, untranslated regions tend to be more variable than coding regions of genes and hence UTRs from different species may be quite dissimilar; therefore even if the main database contains the UTR of an orthologous gene from a closely related species the two sequences may not be significantly similar enough to produce a hit. However, it is much more likely to produce a hit in the EST database, as this contains many more sequences, including many UTRs. By re-screening the EST database using these newly identified, overlapping, sequences it may be possible to produce a longer, composite sequence. Repeated searches of this type may allow you to carry out a 'virtual walk' through the untranslated region into the open reading frame, and this assembled sequence can then be used to BLAST the main database, with a much increased chance of producing a significant hit that contains useful, functional data.

The second major role of EST databases is in the search for polymorphic alleles. As the EST database is derived from sequences from many individuals then it is possible that any polymorphisms in the coding regions of genes from those individuals will be reflected in the database. Initially, the database is searched to identify all the ESTs representing a particular mRNA transcript. These are then aligned using multiple sequence alignment tools such as those at the European Bioinformatics Institute (http://www.ebi.ac.uk/index.html). Any observed differences may represent polymorphisms; however, such information must be confirmed by the analysis of real populations for the presence of these 'virtual polymorphisms'.

8.6.2 Gaining more information about a sequence

We now have the tools to take a DNA or RNA sequence and discover if anything like it exists in the database. However, can we go on to gain more information about it? To better understand the potential roles of this sequence we may wish to ask several questions, including 'What tissues is it expressed in?' and 'Are multiple alleles (polymorphisms) known for this sequence?'.

The first question would traditionally be carried out using a Northern blot, where RNA from several organs is separated on a gel, transferred to a nylon membrane and then probed with the test sequence. However, it is now possible to carry out a 'virtual Northern' using the information available on the WWW. Information on such virtual Northerns is available at the Unigene database, at the following site (http://www.ncbi.nlm.nih.gov/ entrez/query.fcgi?CMD=&DB=unigene). Such a search will tell you from which tissues this mRNA has been identified and submitted to the database. And because of the thoroughness of the genome project it will also tell you, by inference, which tissues it is likely not to be expressed in.

As well as wishing to know what tissues a transcript is expressed in, via virtual Northerns, we may also wish to know if multiple versions of this sequence have been identified. As we have previously seen (Chapters 2, 3 and 7) polymorphisms in drug-metabolizing enzymes may dramatically alter how the body handles a compound, and this may affect the degree of toxicity, if any, that is observed. The UniGene database contains some information on allelic variants, and by comparing the known sequences it would be possible to identify some variants. However, a far more advanced search system is available at databases designed specifically to contain single nucleotide polymorphism (SNP) data. Sites such as the NCBI SNP database (http://www.ncbi. nlm.nih.gov/SNP) and the SNP consortium (http://snp.cshl.org/) provide extensive searching facilities to identify SNPs associated with your genomic DNA sequence of interest, and the DNA surrounding it. A limitation of these databases is that they show where SNPs occur within the target sequence but do not predict what functional effect, if any, these will have on the sequence.

As we will see in the next section, it is also possible to predict protein structure from a DNA sequence. In addition, however, we can gain important information on the regulation of a gene by studying its promoter and enhancer regions. As discussed in Chapter 7, gene expression is under the control of various transcription factors, which bind to specific sites in the promoters and enhancers of genes and affect the rate of transcription. If we know what these sequences are then the next logical progression is to use bioinformatic tools to search promoter regions for putative transcription factor binding sites *in silico*. The most commonly used transcription factor binding site database is TRANSFAC (http://transfac.gbf.de/TRANSFAC) and a number of online search engines can be used to interrogate this database. MatInspector (http://www.genomatix.de/cgi-bin/matinspector/ matinspector.pl) and Alibaba (http://wwwiti.cs.uni-magdeburg.de/~grabe/

alibaba2/) are two of the commonly used engines, and both provide putative transcription factor binding sites within the target sequence, as well as links to provide more information about the factors that bind to these sites. It should be noted however that, as with polymorphism detection via EST comparisons, such assignments are putative, and require 'wet lab' experiments to prove that they are real – just because a site exists does not mean that a factor actually binds there *in vivo*. They do however provide an important 'clue generation' step to target further 'wet lab' experiments.

8.6.3 Proteomics: protein domains, molecular modelling

With the interrogation of the TRANSFAC database we saw a way of predicting promoter functions purely from a DNA sequence. Such an approach can also be taken for the analysis of the coding region of a gene, with the aim of gaining further information on the protein product.

Perhaps the most obvious step is the calculation of primary protein structure from a DNA sequence. Knowing the DNA sequence and the genetic code of the organism it is derived from, it is relatively easy to determine the probable amino acid sequence of the protein product. This sequence can then be used to BLAST a protein database, such as the Protein Databank (http://rcsb.org/pdb/) to search for similar proteins, in the same way as similar DNA sequences are found using BLAST. Such searches will probably highlight multiple different proteins that match different sections of the total sequence, matching to protein domains rather than the whole sequence. Such a strategy may make interpretation of the results difficult as many proteins with seemingly unconnected biological functions may be identified.

The logical extension of this would therefore be to search for the protein domains directly, and infer biological function dependent upon the mixture of domains seen. For example, a protein with ligand-binding, protein–protein interaction and DNA-binding domains is likely to be a transcription factor that interacts with other proteins during its functioning. A number of different WWW resources exist to search for protein domains, and these are given in *Table 8.4*. Of the listed sites InterPro combines many of the other tools into a single unit and hence provides the most comprehensive interrogation tool.

Table 8.4   Software for prediction of protein structure

Site	Internet address
Protein domain prediction software	
PROSITE	http://www.expasy.ch/prosite/
BLOCKS	http://www.blocks.fhcrc.org/
PRINTS	http://www.bioinf.man.ac.uk/dbbbrowser/PRINTS/
InterPro	http://www.ebi.ac.uk/interpro
Secondary structure validation software	
VERIFY3D	http://www.doe-mbi.ucla.edu/verify3d.html
WHATCHECK	http://www.ander.embl-heidelberg.de/whatcheck
AQUA	http://www.nmr.chem.ruu.nl/users/rull/aqua.html
SQUID	http://www.yorvic.york.ac.uk/~oldfield/squid

Data from Fielden *et al.* (2002).

One final piece of information about a protein can be estimated *in silico*, the secondary (3-dimensional) structure of the protein. Protein Databank (*Table 8.4*) contains over 13 000 3-dimensional structures that the sequence of interest can be compared to. Initially, proteins with similar peptide sequences are found, and those for which the 3D structures are also known selected. Proprietary software such as the SYBL suite can then be used to produce a homology model of the 3D structure of the protein of interest, by modelling it on the structures of these similar proteins (homology modelling). This therefore provides the 'best guess' of the 3D structure; to test the validity of this model two *in silico* approaches exist. Firstly, online tools can be used to check if the predicted structure conforms to general rules for protein structures, and is therefore theoretically possible (*Table 8.4*). Secondly, if the protein of interest is known to interact with a ligand, then the ability for this reaction to take place may be tested, as explained in the next section.

8.6.4   Proteomics: compound docking

One key role of bioinformatics in helping to understand the molecular mechanisms of toxicity is aiding in the understanding of how proteins interact with their ligands. This may either be as a substrate binding to the active site of an enzyme, or ligand binding to a transcription factor, leading to changes in gene expression. Computer-based analysis lends itself well to the iterative tasks of trying different models of ligand–protein interactions (i.e. orientation of ligand, key amino acids in protein that make contact with ligand) to determine which is the most energetically favourable, and hence most likely. In addition to modelling known interactions, it should be possible to identify putative new interactions, thus predicting what proteins a novel chemical will interact with, and hence any possible toxicity it may cause.

Ligand–protein interactions can be carried out using several different tools, as described in *Table 8.5*. These tools either use a series of molecular descriptions to specify interaction points, and then build a model around this (descriptor matching) or attempt to dock a ligand into a specified area of a protein without further bias (grid searching). Compound docking is an important tool in prediction of novel interactions, and had been used extensively to predict interactions for both nuclear receptors (e.g. PPARα; Lewis *et al.* 2002a) and enzymes (e.g CYP3A4; Lewis *et al.* 2002b).

Table 8.5   Ligand–protein interaction tools

Program	Method	Website
DOCK	Descriptor matching	http://www.cmpharm.ucsf.edu/kuntz
GOLD	Descriptor matching	http://www.ccdc.cam.ac.uk/prods/gold/gold.html
FlexX	Descriptor matching	http://www.tripos.com/software/flexx.html
AutoDock	Grid-based docking	http://www.cripps.edu/pub/olson-web/doc/autodock

Data from Fielden *et al.* (2002).

8.6.5 Pulling it all together

From the above descriptions it is obvious that there is a tremendous amount of information available on the Internet, and that much analysis of biological functions can be carried out without ever having to enter into a traditional 'wet lab'. One problem with the Internet, however, is that as a direct result of its rapid expansion and the large amount of information available, it can be difficult to find your way and arrive at an integrated answer. While the previous sections detail how to carry out specific tasks, or gain units of information, what happens if I want to know everything I can about a sequence? The first option would be to go through each of the individual sections above, generating genome, transcriptome and proteome information for the sequence of interest. However, a much faster route would be to use a site where such information is collated, and which acts as a 'jumping off point' to other sites with relevant information: using such 'portals' it is therefore possible to quickly gain as much information as possible.

There exist a number of sites that are of use to the bioinformatician to serve as such portals. GeneCards is a database maintained by the Weizmann Bioinformatics institute (http://bioinfo.weizmann.ac.il/cards/). It contains information on *human* sequences, and is fully searchable by name, synonym or accession number. Information provided by GeneCards includes: alternative names, functional roles, sequences available on the databases at both the DNA and protein levels (including the RefSef), SNPs, disease associations, orthologues and the ability to search PubMed for literature related to this gene.

A second portal for useful information is ToxPortal, which contains jumping off points to several toxicology-related sites (http://www.toxportal.com/). Information is sorted by topic, with individual pages for general topics (i.e. biochemistry, statistics, etc.) as well as toxicology (i.e. genotoxicity, carcinogenicity, etc.) and specialist toxicology (i.e. clinical, occupational, etc.).

References

Altschul, S., Gish, W., Miller, W., *et al.* (1990) Basic local alignment search tool. *J. Mol. Biol.* **215**(3): 403–410.

Beckwith-Hall, B.M., Nicholson, J.K., Nicholls, A.W., *et al.* (1998) Nuclear magnetic resonance spectroscopic and principal components analysis investigations into biochemical effects of three model hepatotoxins. *Chem. Res. Toxicol.* **11**(4): 260–272.

El-Sankary, W., Plant, N. and Gibson, G. (2000) Regulation of the CYP3A4 gene by hydrocortisone and xenobiotics: role of the glucocorticoid and pregnane X receptors. *Drug Metabol. Dispos.* **28**(5): 493–496.

El-Sankary, W., Gibson, G.G., Ayrton, A., *et al.* (2001) Use of a reporter gene assay to predict and rank the potency and efficacy of CYP3A4 inducers. *Drug Metab. Dispos.* **29**(11): 1499–1504.

El-Sankary, W., Bombail, V., Gibson, G.G., *et al.* (2002) Glucocorticoid-mediated induction of CYP3A4 is decreased by disruption of a protein:DNA interaction distinct from the pregnane X receptor response element. *Drug Metab. Dispos.* **30**(9): 1029–1034.

Fielden, M.R., Matthews, J.B., Fertuck, K.C., *et al.* (2002) In silico approaches to mechanistic and predictive toxicology: an introduction to bioinformatics for toxicologists. *Crit. Rev. Toxicol.* **32**(2): 67–112.

Gygi, S.P. and Aebersold, R. (2000) Mass spectrometry and proteomics. *Curr. Opin. Chem. Biol.* **4**: 489–494.

Lewis, D.F., Jacobs, M.N., Dickinson, M., *et al.* (2002a) Molecular modelling of the peroxisome proliferator-activated receptor alpha (PPAR alpha) from human, rat and mouse, based on homology with the human PPAR gamma crystal structure. *Toxicol. In vitro* **16**(3): 275–280.

Lewis, D.F., Modi, S. and Dickins, M. (2002b) Structure–activity relationship for human cytochrome P450 substrates and inhibitors. *Drug Metab. Rev.* **34**(1–2): 69–82.

Nicholson, J.K., Connolley, J., Lindon, J.C., *et al.* (2002) Metabonomics: a platform for studying drug toxicity and gene function. *Nature Rev. – Drug Discov.* **1**: 153–161.

Zhang, L., Yu, W., He, T., *et al.* (2002) Contribution of human Defensin 1, 2 and 3 to the anti-HIV-1 activity of CD8 antiviral factor. *Science* **298**: 995–1000.

Index

Milton Keynes UK
Ingram Content Group UK Ltd.
UKHW051935141024
449569UK00027B/1497